DEPARTMENT OF THE

Waste Management Paper No 27

Landfill Gas

A technical memorandum providing guidance on the monitoring and control of landfill gas

London : HMSO

© Crown copyright 1989

Applications for reproduction should be made to HMSO

First published 1989

Second edition 1991

ISBN 0 11 752488 3

It is intended to revise and further update this Waste Management Paper in the future. Any suggestions or enquiries should be addressed to:

The Department of the Environment

Wastes Technical Division

2 Marsham Street

London SW1P 3EB

WASTE MANAGEMENT PAPER No 27
CONTENTS

FOREWORD		**1**

PART 1
SUMMARY

1	**Summary**	**5**
	Landfill gas	5
	Gas movement and migration	5
	Site investigation and assessment	6
	Monitoring	7
	Control	8
	Management	8
	Development on and around landfill sites	9
	Key points for licensing	9

PART 2
LANDFILL GAS EVOLUTION AND CONTROL

2	**Introduction**	**13**

3	**Landfill Gas Characteristics**	**15**
	General Properties	15
	Minor constituents of landfill gas	15
	Flammability	15
	Asphyxiation	16
	Toxicity	16
	Density	16
	Corrosive properties	16

4	**Landfill Gas Generation**	**17**
	Aerobic processes	17
	Anaerobic processes	17
	Factors controlling the processes	18
	Timescale of processes	20

5	**Gas Movement and Migration**	**21**
	Mechanism	21
	Gas pressure	21
	Movement within the site	21
	Migration outside the site	22
	Gas accumulations	24
	Changing patterns of migration	24
6	**Site assessment**	**27**
	Introduction	27
	Closed sites	27
	Operational sites	28
	Obtaining a database	29
	Proposed sites	29
	Other sources of gas	29
7	**Monitoring for landfill gas**	**31**
	Introduction	31
	Monitoring frequency	31
	Weather services for landfill gas management	33
	Supervision of monitoring	34
	Responsibility for monitoring	34
	Review of the monitoring strategy	34
	Emergency monitoring	34
	Trigger values in buildings	35
	Potential for fires on landfill sites	35
	Portable instruments	35
	Gas sample collection	36
	Other aids to monitoring	36
	Monitoring techniques	36
	Surface monitoring	36
	Sub-surface monitoring - gas probes	36
	Sub-surface monitoring - excavated pits and trenches	37
	Sub-surface monitoring - gas monitoring boreholes and wells	37
	Use of leachate wells	41

8	**Control measures for landfill gas**	**43**
	Introduction	43
	Gas barriers	43
	Permeable trenches	44
	Gas drains	45
	Gas wells	45
	Gas pumping systems	47
	Gas plant	49
	Valves and valve selection	50
	Connecting pipework	50
	Maintenance of gas pumping systems	50
	Effects of site operations on gas management systems	51
	Settlement	51
	Safety on sites	51
	Gas management	52
	Commercial utilisation	52
9	**Development on or around landfill sites**	**55**
	Introduction	55
	Development on landfill sites	55
	Development of land adjacent to landfill sites	55
	Highways on or adjacent to landfill sites	56
	Conclusion	56
APPENDIX A	**Typical landfill gas composition**	**57**
APPENDIX B	**Main relevant legislation**	**59**
APPENDIX C	**Monitoring for landfill gas in buildings**	**61**
APPENDIX D	**Description of monitoring equipment**	**67**
APPENDIX E	**Sample collection and measurement techniques**	**71**
APPENDIX F	**Other aids in gas monitoring and measurement**	**73**
APPENDIX G	**Gas measurements and sampling at boreholes and gas wells**	**75**

BIBLIOGRAPHY		79
GLOSSARY OF TERMS		81
FIGURE 4.1	Decomposition pathway of materials occurring in household waste	18
5.1	Gas movement within wastes	22
5.2	Possible gas migration paths from a completed/restored site	23
7.1	Sub-surface probe (steel pipe)	38
7.2	Sub-surface probe (plastic pipe)	38
7.3	Simple monitoring borehole	39
7.4	Multiple point monitoring borehole	39
8.1	Permeable trench venting system	46
8.2	Passive gas venting well	47
8.3	Typical landfill gas extraction well	48

FOREWORD

Landfill gas is an end product of the degradation of biodegradable wastes in a landfill site. Typically it is a mixture of up to 65% methane and 35% carbon dioxide plus trace concentrations of a range of organic gases and vapours. Methane is flammable at concentrations between 5 and 15% by volume in air. Without proper management the migration of gas from a landfill can give rise to risk of fire and explosion in nearby buildings, underground services or voids. It also presents a risk of asphyxiation.

This Waste Management Paper revises and adds to the first edition and should assist all those concerned in dealing with landfill gas. It is, however, recognised that knowledge of many aspects of landfill gas management is still under development. Accordingly this revision continues the precautionary approach adopted previously. Consequently all parties involved in landfill management are urged to take care in specifying the types and the combinations of controls to be employed. Therefore, although guidance is given on the factors to be taken into account in setting licence conditions, under either the Control of Pollution Act 1974 or in due course, the Environmental Protection Act 1990, standard licence conditions should not be applied. Each site should be treated on an individual basis, though in general, single systems of control are unlikely to be adequate at most landfills.

Part 1 is a summary of the Waste Management Paper. Part 2 is a discussion of the main factors responsible for the formation of landfill gas. It describes the properties of the gas mixture, factors affecting its migration, and methods of assessing, monitoring and controlling the gas, in order to give information on the options available for its management. The Paper does not consider the management and control of leachate, even though it is recognized to be closely associated with gas.

This revised Waste Management Paper was prepared with the assistance of the following representatives of organisations with an interest in waste management:

Chairman

Mr J Nicholson   HM Inspectorate of Pollution

Members

Mr H J Barrie   Durham County Council (representing
                              Association of County Councils)

Mr C A R Biddle   Representing Road Haulage Association

Mr K J Bratley	West Yorkshire Waste Management (representing Association of Metropolitan Authorities)
Mr D J V Campbell	Harwell Laboratory
Mr M Crawford	Scottish Office Environment Department
Mr C J Cooke	Blue Circle Industries PLC (representing County Surveyors Society)
Mr S Davies	South West Wales Waste Management Group (representing Association of District Councils)
Mr J Gunstone	Scottish Office Environment Department
Mrs T Hillman	Representing Institute of Wastes Management
Mr H D T Moss	Shanks and McEwan Ltd
Mr F Mulgrew	City of Dundee (representing Convention of Scottish Local Authorities)
Mr C Palmer	Suffolk County Council (representing County Surveyors Society)
Mr A Sheppard	Tarmac Econowaste Ltd (representing National Association of Waste Disposal Contractors)
Mr A J Smith	Department of the Environment
Mr G D Small	Representing Confederation of British Industry
Mr R Tomlinson	Representing Waste Disposal Engineers Association
Mr R V Watkinson	HM Inspectorate of Pollution/Department of the Environment
Mr C Welch	West Yorkshire Waste Authority (representing Association of Metropolitan Authorities)
Mr H O'Connor	HM Inspectorate of Pollution (Secretary)

The following also attended meetings and contributed to the preparation of this revised Waste Management Paper:

Ms C John	HM Inspectorate of Pollution
Mr G Baldwin	Harwell Laboratory
Mr S Gibbs	Acer Environmental

The Health and Safety Executive

In this Waste Management Paper, the term "Waste Regulation Authority" (WRA) is used whenever reference is made to the regulatory function of Waste Disposal Authorities under the Control of Pollution Act 1974 or the Waste Regulation Authority under the Environmental Protection Act 1990.

PART 1
SUMMARY

CHAPTER 1

Summary

Landfill gas

1.1 Wherever biodegradable material is deposited in landfill sites, microbial activity will generate landfill gas, which is a mixture of flammable and asphyxiating gases. It therefore follows that all sites should be assessed, monitored and, where necessary, have control systems installed to prevent uncontrolled gas migration.

1.2 Biodegradable material is any organic matter which can be decomposed by microorganisms. This will include, for example, animal and vegetable matter, paper and wood (including timber products, shrubs and trees). Many classes of controlled waste contain these materials in varying proportions and they can be expected to be found in household, commercial and industrial wastes and in the categories of materials known as builders' waste, "civic amenity" or bulky household waste and garden waste. Much waste classed as "inert" contains wood and paper. (Para 2.2)

1.3 Landfill gas is evolved from the breakdown of biodegradable wastes in a landfill. The composition of the gas varies according to the type and phase of breakdown which is occurring within the site at any specific time. Initially carbon dioxide predominates, though significant quantities of hydrogen are also evolved. Methane (about 65%) and carbon dioxide (about 35%) are the major constituents of the gas evolved during the usually predominant anaerobic phase of waste breakdown. The onset and rate of degradation processes in the wastes vary both within and between sites. The evolution of significant quantities of methane may take from three months to more than a year to start and can continue for well in excess of 15 years. (Paras 4.1–4.5, 4.17)

1.4 There are many factors which influence gas evolution including the physical dimensions of the site, the types of waste and their input rates, moisture content, landfill pH, temperature and waste density, together with site operational practice. (Paras 4.6–4.16)

1.5 The major constituents of landfill gas are colourless and odourless although they are normally found mixed with other gases, some of which give rise to odours. It is usually saturated with moisture, and is corrosive. The density of landfill gas is dependent on the relative proportions of its major components, but is usually about the same as air. (Para 3.1)

1.6 Gas pressure within a landfill is dependent on the gas evolution rate, the permeability of the fill, and the permeability of the surrounding strata; it can be varied by changes in the level of the leachate in the site. Differences in atmospheric pressure will affect the pressure differential between the site and the atmosphere which in turn affects gas emissions from the site. (Paras. 5.1–5.11)

1.7 If not properly monitored and controlled, landfill gas can give rise to flammability, toxicity and asphyxiation hazards as well as vegetation dieback. (Paras 2.1, 3.2–3.5)

Gas movement and migration

1.8 Gas may move in any direction within the wastes. Lateral gas movement will be encouraged by low permeability compacted layers whilst vertical gas movement will occur around gas and leachate wells particularly if they have been surrounded by hardcore. Gas may also move vertically at the sides of a site at the interface between the wastes and surrounding strata and escape via settlement cracks. (Paras 5.3–5.5)

1.9 Gas migration from the site can occur in several ways. It will move through permeable strata or for considerable distances along faults, fissures or cavities in the strata. It can pass along man-made features such as mine shafts, roadways, sewers, or along the backfill around pipes or cableways. Gas may also dissolve in leachate or groundwater or be evolved from leachate degradation and subsequently released some distance from the site boundary. (Paras 5.6–5.11)

1.10 Migration pathways are affected by surface sealing which may be caused by heavy rainfall, ice or snow, by changes in the permeability of the waste as it settles and decomposes or by subsequent disturbance of the site. Gas movement also alters with variations in gas and atmospheric pressure. Emission of landfill gas may be detectable by smell, sound, or by the presence of bubbles in surface water. Vegetation may also be adversely affected by landfill gas in the ground resulting in bare patches or brown foliage and subsequent dieback. (Paras 2.1, 5.13–5.15)

Site investigation and assessment

Closed sites

1.11 All land that contains controlled wastes should be assessed for the potential for landfill gas evolution. This assessment should include: a desk study detailing the geology, hydrogeology, topography around the site; waste types, age and depths; and development on and within at least 250 metres from the boundary of the landfill. Where uncontrolled gas migration is found and the quantities of gas evolution at the site are liable to create a hazard, a gas control system should be installed or, if one exists, it should be improved. (Paras 6.3–6.10)

1.12 The responsibility for the assessment and monitoring of closed sites for which there is no licence rests primarily on the land-owner. The Environmental Health Authority should also assess the likely risk to public health or of nuisance and decide in the light of evidence available what action should be taken. (Para 7.15)

1.13 Under Section 61 of the Environmental Protection Act 1990, Waste Regulation Authorities (WRAs) will have the duty to inspect the land within their area to detect whether any gas is being emitted or discharged from any former landfill site. Where significant quantities of gas are being evolved at any site without a waste management licence and are likely to cause harm to human health or pollution of the environment, then the WRA has the duty to take action to ensure the gas is properly controlled. (Para 7.16)

Operational sites

1.14 All site operators should undertake a site assessment, as described in Chapter 6. This should be agreed with the WRA. The location of monitoring boreholes and wells and any other action deemed necessary should form part of the working plan and licence for the site. (Paras 6.13, 7.6)

Proposed sites

1.15 Potential operators of proposed sites should undertake an assessment of the geology surrounding a proposed site. The gas management scheme should be included within the working plan of the site where appropriate. (Paras 6.15, 6.16)

1.16 No landfill site should be authorised by licensing until it has been comprehensively assessed and a satisfactory gas management scheme included in the site working plan. (Paras 6.15, 8.1)

Monitoring

1.17 A programme of monitoring for landfill gas should be established at all sites that have taken controlled wastes, following the site assessment. For proposed sites it should commence prior to any deposit of wastes and for all sites should continue until the biodegradation process has ceased. Monitoring should continue until either:

a) the maximum concentration of flammable gas from the landfill gas remains less than 1% by volume and the concentration of carbon dioxide from the landfill remains less than 1.5% by volume measured in all monitoring points within the wastes over a 24 month period taken on at least four separate occasions, including two occasions when atmospheric pressure was falling and was below 1,000 mb; or

b) an examination of the waste using an appropriate statistical sampling method provides a 95% level of confidence that the biodegradable matter has been used up. (Paras 7.1, 7.9, 7.10)

Care should be taken to distinguish landfill gas from other sources of gas. (Paras 6.17–6.20)

1.18 Monitoring should be carried out by a competent trained person and records made and kept. A standard method needs to be adopted so that results are comparable, reliable and reproducible. (Paras 7.14 & Appendix G)

1.19 Portable instruments used to detect major components of landfill gas should be carefully selected and regularly calibrated. The results obtained should be confirmed periodically (at least twice yearly) by gas chromatography. Great care should be exercised when collecting samples for analysis. (Paras 7.23, 7.24, Appendices D & E)

1.20 A number of techniques for surface and sub-surface monitoring are available. The preferred method for sub-surface measurements of gas is by monitoring boreholes, which should reach below the depth of the site or into the permanent water table, and by gas extraction wells intercepting all levels of wastes. (Paras 7.28–7.48)

1.21 Borehole spacing around a site, and positions of gas extraction wells, will be site specific. They will depend on the volumes of gas being evolved, geology of the site, development around the site and effects on vegetation. (Para 7.45, Table 7.1)

1.22 Whatever type of site is being monitored, the monitoring strategy should be to establish what is occurring both in and around the site in relation to gas evolution and migration. It should establish whether gas is escaping from a site in an unplanned or uncontrolled manner and whether it is likely to become a risk to public health or the environment or a nuisance. The information gained by monitoring should be processed to update the assessment being undertaken of the gas management scheme at the site. (Para 7.1)

1.23 Monitoring frequencies are site specific and dependent upon the assessment of risk to people and development from gas migration. They should be regularly reviewed, as part of the site assessment. After obtaining a database, monitoring frequencies may vary from twice per annum for isolated sites to weekly for urban sites. Continuous monitoring may be necessary if persons or property are at risk. (Paras 7.4–7.12)

1.24 Emergency procedures should be established by operators, Waste Regulation and Environmental Health Authorities. The Emergency Services, the Health and Safety Executive, British Gas plc and, where relevant, British Coal should be consulted when drawing up these procedures. (Para 7.20 & Appendix C)

Control

1.25 Gas should not be allowed to escape from a landfill in an unplanned or uncontrolled manner. It should be contained and vented within the perimeter of the site or flared in a secure compound, designed for that purpose. All control systems should have adequate protection against failure. (Paras 8.1 & 8.2)

1.26 In general, gas control measures should be implemented or improved if, in any subsurface probe or borehole outside the area of influence of the gas control system or the area of the wastes, whichever is greater, the concentration of flammable gas from the landfill gas exceeds 1% by volume and the concentration of carbon dioxide, from the landfill, exceeds 1.5% by volume. It is essential to distinguish landfill gas from other sources. (Paras 2.4, 6.17–6.20)

1.27 There are various methods of control available. These include gas barriers, venting trenches and wells. The effectiveness of wells can be improved by the use of gas drains. The sphere of influence of trenches and wells can be increased by pumping. (Paras 8.4–8.25)

1.28 When sites require gas pumping systems the collected gas should normally be flared off or preferably utilized. When it cannot be flared it may be vented in a safe manner. The design and installation of systems requires specialist advice. After installation, systems require regular maintenance and may need readily available replacement plant. (Paras 8.26–33, 8.39)

Management

1.29 Site operators should ensure the gas management scheme is not compromised by waste disposal operations and that the risks from landfill gas are minimized. Measures should include:–

- Prevention of perched water within wastes;

- Careful siting of temporary roads;

- No ignition sources on site (including no smoking);

- Careful use of vehicles;

- Regular monitoring of site buildings for landfill gas;

- All electrical equipment to comply with the appropriate British Standard, and

- Adequate safety instructions to all site personnel.

(Paras 8.41–8.51)

1.30 The various control and monitoring strategies should provide an effective, permanent gas management system. This is likely to require the installation of more than one means of gas control and should be designed with specialist advice. Where gas is used as an energy source, special care is necessary when gas is not being utilized. (Paras 8.52–8.56)

Development on and around landfill sites

1.31 After-use of landfill sites should normally be restricted to agriculture or similar uses. Where control systems permit, the land may be used for public open space, conservation or recreation. It is recognized that there will be pressures for other forms of development. This should be possible provided adequate precautions are taken. In particular, older landfills in urban areas can be developed for non-housing uses such as supermarkets and light industrial units subject to detailed investigation and design in accordance with current advice and guidance. It is recommended that no housing should be built within 50 metres of any landfill site with gas concentrations in excess of the limits given in paragraph 7.9 or which still have the potential to produce large quantities of gas. Gardens of houses should not extend to within 10 metres of any such site. Great care should be taken with any development which takes place within 250 metres of in-filled wastes. Where development is taking place on or adjacent to a landfill site, the developer should take account of the need for the assessment and monitoring of any risk to the development posed by the site. (Paras 9.1–9.9)

Key points for licensing

1.32 Waste Regulation Authorities should relate licence conditions with respect to landfill gas to the specific requirements of the site. The general statements contained in this paper should not be considered to be model licence conditions. The following points are listed for assistance when drawing up licences:–

a) All sites should be considered as having the potential to generate landfill gas;

b) All sites should be monitored for landfill gas evolution and migration;

c) A desk study should be undertaken to identify potential risks at and around all sites;

d) A thorough site survey of the type described in Waste Management Paper No. 26 Paragraphs 3.109–3.119 should be carried out, covering an area of at least 250 metres from the boundary of the site and all areas identified by c) above;

e) Site design should include the design for the gas control system in the working plan;

f) Site operations should not compromise the gas control system;

g) Gas can move in all directions. The pattern of such movement can vary with time;

h) Gas monitoring should be undertaken by trained personnel using properly calibrated equipment;

i) All monitoring should be carried out in accordance with the site working plan, which should detail the monitoring regime for the site;

j) Records of monitoring should be made and retained in a clearly understandable format;

k) Regular inspection and maintenance of the gas management systems should be carried out;

l) Repairs to the gas management system should be carried out promptly;

m) Emergency procedures should be drawn up between the site operator and the WRA;

n) Emergency procedures should be included with the working plan;

o) Monitoring of sites should continue and no waste management licence accepted for surrender until the conditions described in Para 7.9 have been fulfilled;

p) No licence should be issued until the WRA is satisfied with the gas monitoring and, where necessary, gas control arrangements, as detailed in the working plan;

q) It is likely that many sites will require a combination of gas control measures; and

r) Monitoring points should be established between the landfill site and any area at risk from uncontrolled gas migration.

PART 2
LANDFILL GAS EVOLUTION AND CONTROL

CHAPTER 2

Introduction

2.1 An increasing number of incidents and growing public concern have focused attention on the fate of the end-products of waste degradation in landfill sites and their effects on the environment. One major area of concern is the evolution of landfill gas which, when inadequately controlled, has led to explosions, fires, dangerous gas concentrations in and around houses, odour nuisance and vegetation dieback. This Waste Management Paper provides guidance for WRAs, site operators, site owners, and local authorities and may also be of interest to consultants, property developers, and others who may be involved in the management of landfill gas.

2.2 The evolution of landfill gas is an inevitable consequence of the disposal of any biodegradable wastes in a landfill. (Biodegradation is the breakdown of substances by the action of micro-organisms.) Biodegradable wastes include all putrescible wastes, both animal and vegetable matter, paper, and any type of wood such as shrubs and trees, including timber products. Microbial activity during the degradation process produces a mixture of gases consisting principally of methane and carbon dioxide. The types of waste from which landfill gas is produced are household and commercial in origin but will include some types of industrial wastes. Much waste classed as "inert" has contained wood and paper. Rates of gas evolution depend on the physical, chemical and microbial characteristics of the landfill. Such variables also affect the timescale of gas evolution.

2.3 For proposed and operational sites, the position of monitoring points and the need for the installation of a gas control system should be evaluated following a detailed assessment of the site and the surrounding area. Restored sites and those awaiting restoration should be assessed to establish the composition of the gas, its rate of evolution, any off-site migration routes and the extent of any potential hazard. From this information the extent of monitoring can be established and a suitable gas control system installed where necessary.

2.4 The aim of any gas control system must be to prevent uncontrolled gas migration; the objective being to ensure that landfill gas does not pose a risk to human health or pollution of the environment. A control system should be considered effective if the concentration of flammable gas from the landfill never exceeds 1% by volume and the concentration of carbon dioxide, from the landfill, never exceeds 1.5% by volume in any sub-surface probe or borehole outside the area of influence of the gas control system or the area of the wastes, whichever is greater. The sphere of influence of a gas control system will normally be wholly within the deposited wastes. It is essential to distinguish landfill gas from other sources of gas. It may be necessary to make an allowance for existing background levels. On occasions where control measures have been installed outside the area of the waste deposits, the land between the wastes and the control measures should be treated as if it contained wastes. (see Paragraphs 6.15–6.20).

2.5 Where development is proposed or occurs within 250 metres of the boundary of the site, then specialist advice should be sought as to what measures are required to ensure the safety of such development. Attention should be given to the geology of the site, its location in relation to the proposed development, the age and type of wastes deposited, the kind of development involved and any other local factors, such as the presence of other sources of gas (eg mine gas etc.).

2.6 The recovery and utilization of landfill gas as an energy resource may be a commercially viable proposition and can offset some of the costs of control. Although this will contribute substantially to minimising uncontrolled gas emissions and migration, it should not be considered as a substitute for gas control. Additional information is likely to be necessary both to establish the commercial viability of the gas use scheme, and to determine its impact on operational or proposed gas migration control measures.

2.7 While information on the current state of the art is given in this document, it should be used in a cautious and precautionary manner. Research is continuing and practical experience is likely to provide further refinement to the guidance presented. Operators and WRAs should ensure the installation of suitable and appropriate gas management systems, bearing in mind the nature of wastes deposited and assessment of the risks. Licences relating to landfill sites taking controlled wastes should be formulated to accord with this principle.

CHAPTER 3
Landfill Gas Characteristics

General Properties

3.1 The usual major constituents of landfill gas are methane and carbon dioxide both of which are colourless and odourless (the presence of trace components may or may not impart an odour). Landfill gas containing the flammable gases methane and hydrogen can form flammable mixtures in air. It may also act as an asphyxiant either alone or when mixed with air, when the oxygen content is depleted. Carbon dioxide will be a hazard to health if present in concentrations in excess of 1.5% by volume. Undiluted landfill gas can be expected to have a calorific value of 15 to 21 MJm^{-3} (mega- joules per cubic metre), about half that of natural gas, and will have a density close to that of air. Landfill gas will usually be corrosive, saturated with water vapour and will normally be above ambient temperatures. A typical analysis of the gas is given in Appendix A.

Minor constituents of landfill gas

3.2 Many minor constituents are present in landfill gas at low concentrations. Considerable variations in the concentration and presence of trace compounds have been observed, related to waste constituents, age and extent of waste degradation. Organosulphur compounds and esters are found in gases derived from recently deposited wastes, from which odours are more obvious. Trace gases are evolved by the complex interaction of the physical, chemical and biological processes going on within the waste. Some components may need to be diluted more than one hundred million times to be below odour threshold values. The impact of some trace constituents on gas control or utilization equipment may be significant. Volatile aromatic compounds and mercaptans are often responsible for landfill gas odour but hydrogen sulphide, often erroneously thought to be responsible, is generally of low concentration and is normally not a significant health hazard. However hydrogen sulphide can be evolved in quantity from the co-deposit of sulphate- containing materials, such as wastes containing gypsum from plasterboard or some colliery spoils, with biodegradable wastes.

Flammability

3.3 When methane or hydrogen is mixed with air, within certain concentration limits known as the "flammable" or "explosive" range, the resultant mixtures may ignite to produce fires and explosions. The concentration limits are commonly known as the "Lower Explosive Limit" (LEL) and the "Upper Explosive Limit" (UEL), and sometimes referred to as the "Lower Flammable Limit" and "Upper Flammable Limit" respectively. The flammable ranges of methane and hydrogen are 5–15% and 4–74% by volume respectively. The presence of carbon dioxide affects these ranges, but there is little change at the lower limit. When the oxygen concentration falls to below 13% by volume, under normal conditions methane cannot ignite. Flammable mixtures with air can develop from landfill gas evolved during the mature phase of waste biodegradation. When such mixtures accumulate and are ignited within a confined space eg a building, cavity or underground chamber, an explosion may result. In unconfined situations such as the open air, flash fires may result.

Asphyxiation

3.4 An asphyxiation risk from landfill gas is present whenever persons have to enter any confined space in or near a landfill site. This may include manholes, sewers, or tunnels and even poorly ventilated spaces such as may be found below portable buildings etc. No one should enter or remain in any confined space where the oxygen content of air has fallen below 18% by volume as specified in the Health and Safety Executive Guidance Note EH 40 entitled "Occupational Exposure Limits" which is issued annually. This Guidance Note provides a Short Term Exposure Limit for carbon dioxide of 1.5% by volume over 10 minutes with an occupational exposure standard of 0.5% by volume over 8 hours. Elevated levels of carbon dioxide affect human respiration and can prove fatal.

Toxicity

3.5 Some of the minor constituents of landfill gas could have toxic effects if present in high enough concentrations. Operators should undertake an assessment of risk and where necessary apply control measures as required by the Control of Substances Hazardous to Health Regulations 1988. Trace gases do not usually represent a health hazard following normal atmospheric dilution . Hydrogen sulphide is toxic at low concentrations, having occupational exposure standards of 10 ppm (8 hour Time Weighted Average reference period) and 15 ppm (short term exposure, 10 minute reference period) (HSE Guidance Note EH 40).

Density

3.6 The density of landfill gas depends on the proportion of components present. Thus a mixture of 10% hydrogen (density 0.08 kgm^{-3}) and 90% carbon dioxide (density 1.98 kgm^{-3}) typically evolved in the early stages of anaerobic degradation will be denser than air (density 1.29), while 60% methane (density 0.72 kgm^{-3}) with 40% carbon dioxide will be slightly lighter than air. Landfill gas components do not normally separate when collecting in voids but layers of landfill gas may form in still air conditions as a result of density differences. The evolution of warm landfill gas gives it bouyancy in colder air.

Corrosive properties

3.7 Some of the components of landfill gas, or derivatives thereof, have a corrosive potential. Carbon dioxide is soluble in water and will give rise to an ionised solution of carbonic acid in aqueous condensates associated with the gas. Such condensates can corrode a range of metals. The combustion of landfill gas in utilisation or control systems will give rise to a high temperature environment in which some trace components, such as halogenated or sulphuretted compounds, may give rise to highly acidic and aggressive derivatives. At such elevated temperatures, carbon dioxide, hydrogen, hydrocarbons and water vapour can be involved in decarbonisation (removal of carbon from an alloy) or carburization (coking) reactions with alloys strengthened by the addition of interstitial carbon. The engineering and design of landfill gas control systems should address the corrosive potential of landfill gas.

CHAPTER 4
Landfill Gas Generation

Introduction

4.1 The microbial decomposition processes occurring in landfill sites are complex and not yet fully understood. Biodegradable matter in wastes can undergo two forms of decomposition involving aerobic (in the presence of oxygen) and anaerobic (absence of free oxygen) micro-organisms.

Aerobic processes

4.2 Aerobic decomposition predominates during waste deposition and continues until there is insufficient free oxygen to sustain the process. It is therefore possible for the aerobic process to continue until decomposition of the waste is complete. This is rare and generally only takes place at very shallow, uncovered landfills or more frequently for a limited period in uncovered waste in the top layer of the landfill.

4.3 The major decomposition products of this process are water and carbon dioxide. The process is exothermic, which may accelerate the onset of anaerobic conditions. The characteristic odour often associated with aerobic waste degradation is mainly due to the presence of organic esters.

Anaerobic processes

4.4 The anaerobic decomposition of organic matter in wastes takes place in several stages, represented in Figure 4.1. Each stage has an impact on the quality and rate of gas evolution before eventual stabilization. Initially wastes are hydrolysed and fermented by microbes to break down complex organic molecules with evolution of hydrogen and carbon dioxide gases. Further breakdown of fatty acids and alcohols provides substrates for methanogenic bacteria finally to convert either acetate to methane and carbon dioxide or utilize hydrogen and carbon dioxide to form methane and water. In addition microbial, chemical and physical processes taking place within the wastes produce many other trace components in landfill gas mixtures. Under the steady state conditions of a normal landfill environment, each stage of waste degradation occurs concurrently; methane and carbon dioxide gases form the bulk of the mixture of gas evolution in the ratio of about 3:2. Landfill gas quality and evolution rates will be dependent on many site specific factors, with significant gas production likely for several decades before a semi-aerobic and ultimately aerobic environment in remaining stabilized wastes becomes re-established.

4.5 During the initial phases of anaerobic degradation landfill leachates often have a high Biochemical Oxygen Demand (BOD) and high Chemical Oxygen Demand (COD) with BOD/COD ratios of around 0.7. The presence of industrial wastes can influence this ratio. As anaerobic degradation proceeds the ratio reduces to about 0.1 or even lower as the degradable components are used up, partly, in the formation of landfill gas. It follows that leachate composition can be indicative of the progress of waste degradation and gas evolution potential. It is important to note that during the time that waste is being deposited, and probably for some period thereafter, gas and leachate compositions will reflect various stages of both aerobic and anaerobic waste degradation.

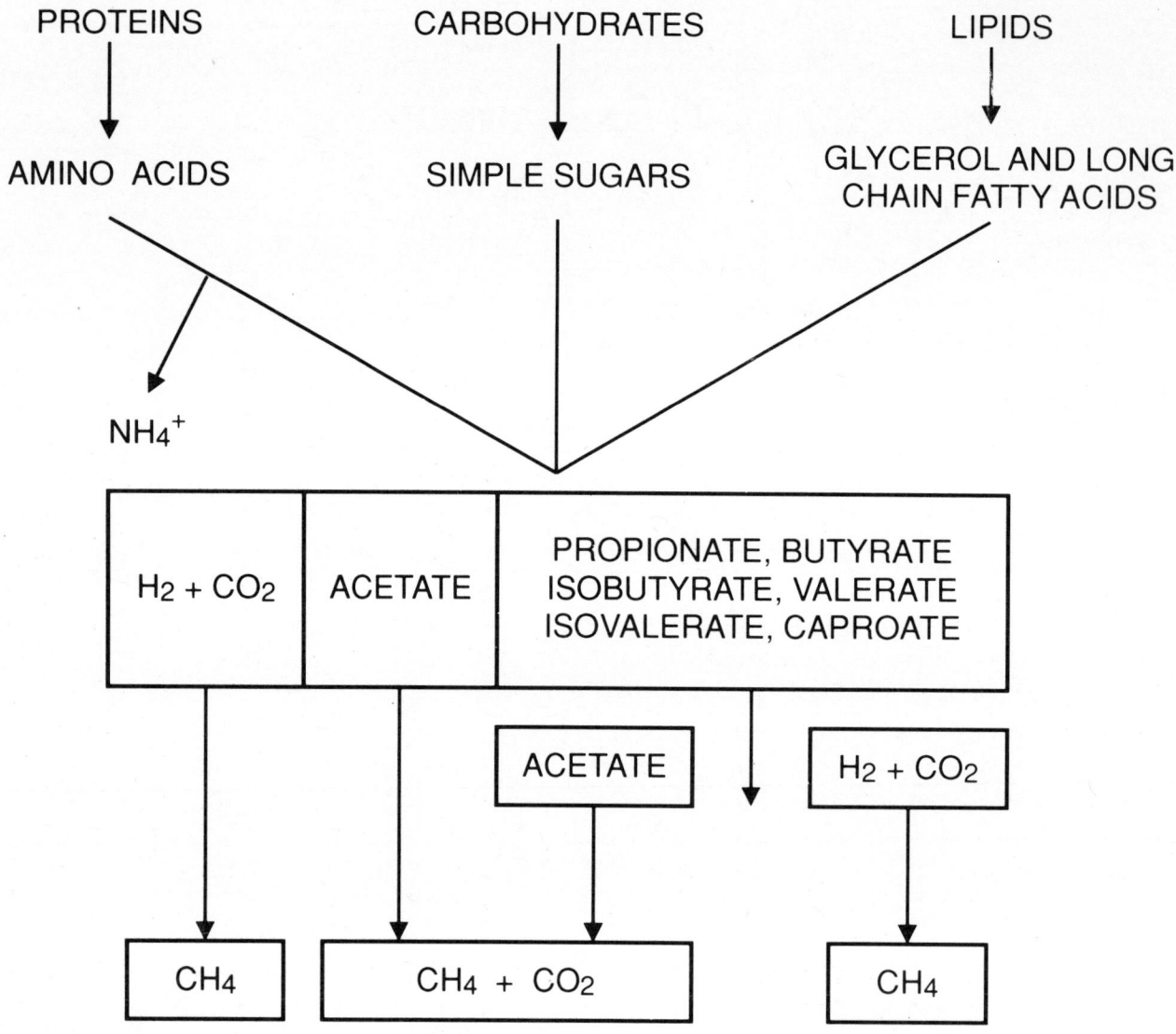

Figure 4.1 Decomposition pathway of materials occurring in household waste

Factors controlling the processes

4.6 The important factors controlling the rates and amounts of gas likely to be evolved are as follows.

Physical dimension of the site

4.7 Anaerobic processes normally dominate in wastes at depths greater than 5 metres. Similarly, these mechanisms are also likely to be occurring in shallower, well-capped sites. Much will depend on the activity within the site, its topography and the opportunity for gaseous exchange at the site surface.

Waste type

4.8 The composition of waste affects the rate, quality, and quantity of gas generated per unit mass. Initial gas composition may derive from the more readily degradable organic matter, while some components of the waste such as heavy metals may, at least locally, inhibit gas generated. Different organic materials in deposited wastes will produce slight differences in methane/carbon dioxide ratios. Mixed waste streams may also react within the site to produce other gases such as hydrogen sulphide.

Waste input rates

4.9 Progressive restoration coupled with high input rates will encourage more rapid development of the anaerobic process.

Site operations

4.10 Reduction of the particle size of the waste, by pulverisation and compaction using thin layering techniques, will hasten the onset of anaerobic decomposition for the more readily degradable materials. Rapid infilling of small areas of a site will shorten the aerobic degradation phase and tend to keep waste temperatures down. This method will also reduce rainfall infiltration. The net result will be slower initial rates of waste degradation. Where large volumes of high BOD leachate are produced and removed from the sites, the resultant loss of nutrients on which landfill gas evolution relies will reduce the overall quantities of gas produced from the site. Daily or intermediate cover and the use of low permeability materials in cell wall construction may encourage perched water tables to develop and have effects on moisture movement, transmission of gases and buffering of leachates. Such effects will be important in terms of gas evolution, migration pathways and proposed methods of gas control.

Waste density

4.11 The degree of saturation of the waste and its density depends on both the void space and the absorptive capacity of the waste. The greater the waste density in a landfill the higher the theoretical yield of landfill gas per unit volume of void space. However, water movement within the waste is necessary to permit the free movement of nutrients for bacteria to flourish. High waste densities will also serve to reduce permeability of the waste to gas and hence result in a build-up of gas pressure.

Moisture content

4.12 A moist landfill environment is normally associated with high rates of gas generation. There are examples of sites whose gas evolution is substantial even though the degree of saturation is apparently low. Incoming household waste has an average moisture content of about 25%, food and garden waste providing the highest moisture input. Thereafter rainfall, surface and groundwater infiltration and the products of waste breakdown can provide additional moisture. The recirculation of leachate practised on some sites will maintain high moisture contents and may provide a source of nutrients and bacteria which will tend to accelerate gas generation rates. Liquid movement within wastes tend to provide a more even waste moisture content. It also distributes nutrients and bacteria within the mass which can further enhance rates of waste degradation and gas evolution. Extraction of the gas itself can assist this process by drawing moist gases through the fill.

pH within the landfill

4.13 Methanogenesis will proceed optimally between a pH range of 6.5 to 8.5 and is only inhibited when the pH value is outside this range. In particular, household waste produces acidic leachate as a consequence of rapid degradation of easily biodegradable material. Unless this is buffered by other wastes it may be responsible for inhibiting the onset of methane evolution. Thus waste streams containing a mix of both biodegradable and "inert" material are more likely to develop a pH in the optimum range.

Waste temperature

4.14 The optimum temperature range for maximising methane production is between 35 and 45 degrees Celsius, which is common in deep landfill sites. A significant reduction in gas evolution occurs below 10 to 15 degrees Celsius. In shallow landfill sites variations in gas evolution rates may among other factors reflect seasonal changes in ambient temperatures.

Ingress of oxygen

4.15 Ingress of oxygen into anaerobically decomposing wastes can occur by excessive pumping rates in landfill gas extraction schemes, or through trenches dug into mature wastes for site operations. Any such ingress will stop the anaerobic phase, and will lead to increases in temperature by the reintroduction of the more exothermic aerobic processes. If this process is continued there is a risk of an underground fire. However, when the air ingress is halted, the anaerobic phase is usually re-established fairly quickly.

Interactions of factors

4.16 All the above factors interact with one another in various ways to influence evolution of gas as well as gas movement and migration. For example, high waste input rates may be wanted to develop anaerobic conditions for gas utilization. This could be countered by effective water management, to control leachate generation. The concentrated production of organic acids in the landfill environment may, as a result, lower the pH, hence inhibiting gas generation.

Timescale of processes

4.17 The diverse nature of wastes and the variability of landfill sites and operational practice makes it difficult to predict the time of onset and duration of gas evolution. Aerobic processes rely on availability of oxygen from the atmosphere. This can be significantly affected by the factors described above. With good compaction and the use of intermediate (daily) cover material it can be expected that aerobic processes will decline within a few days. Thereafter anaerobic processes will predominate and substantial amounts of methane can be expected to be evolved within 3 to 12 months of waste deposition. The concentration of methane will gradually increase until it reaches a typical maximum of between 60 to 65% by volume of the landfill gas. Methane and carbon dioxide may continue to be produced over several decades. Gas generation rates can be expected to plateau and then decline at a rate depending on site conditions. There is no typical figure for the length of time that landfill gas will be evolved but at many sites significant gas generation can be expected to continue for a period which is likely to be at least 15 years after the last deposit of waste. Gas generation may be reactivated; for example if development occurs on the site or liquid levels within the wastes are allowed to rise when leachate pumping is stopped.

CHAPTER 5
Gas Movement and Migration

Mechanism

5.1 The movement of gas within the site and migration out of it are governed by a number of factors which will be site specific. The underlying mechanisms responsible for the migration of landfill gas are provided by the pressure developed within the site, the pressure gradient within the surrounding strata and the concentration gradients of the component gases. The extent of gas movement within and beyond site boundaries will be governed by these mechanisms and the relative permeability of the wastes and surrounding strata.

Gas pressure

5.2 Gas pressure generated within the landfill is dependent on the microbial activity in the waste, its permeability to gas and the permeability of cover and bunding material. The pressures obtained can in theory be high, although this is unlikely to be a significant hazard in itself. An increase in gas pressure will promote migration from the site. This may be aided by:–

* Changes in atmospheric pressure;

* Changes in leachate levels in the waste; and

* Changes in the water table outside the site.

Any combination of the above may alter the flow rates and pathways of the gas as it moves from areas of higher to lower pressure.

Movement within the site

5.3 Gas movement within the site is influenced by a number of factors, the most important of these being the method of filling the landfill site. The combination of a thin layering method of filling with heavy compaction and the use of daily cover materials, particularly where these are of low gas permeability, will help to promote horizontal gas flow. The use of such cover materials will also lead to the development of perched water tables which in turn will further reduce vertical gas flow. Figure 5.1 illustrates how these factors affect gas movement within the wastes. However gas will move vertically through or around boreholes, leachate or gas chimneys and other pathways (eg site roads) installed within the landfill, unless they are sealed. This movement will be further encouraged if wastes, such as builders' rubble, have been placed around the chimneys to protect them from damage or in "christmas tree" fashion to improve the effectiveness of the well for gas extraction purposes. Natural movement paths will also develop in the waste as settlement occurs. Where cellular methods of landfilling are adopted then bunds around cells will limit the lateral migration. This will give rise to surface emissions at the cell waste/wall boundaries until final capping occurs. Gas movement within baled waste sites will initially be via the interfaces between the bales.

5.4 It is good landfill practice to construct a low permeability cap, which can be enhanced by doming or sloping, above the final layer of the waste, to reduce surface water infiltration into the site. However, if steps are not taken to vent the landfill gas, capping will inevitably result in

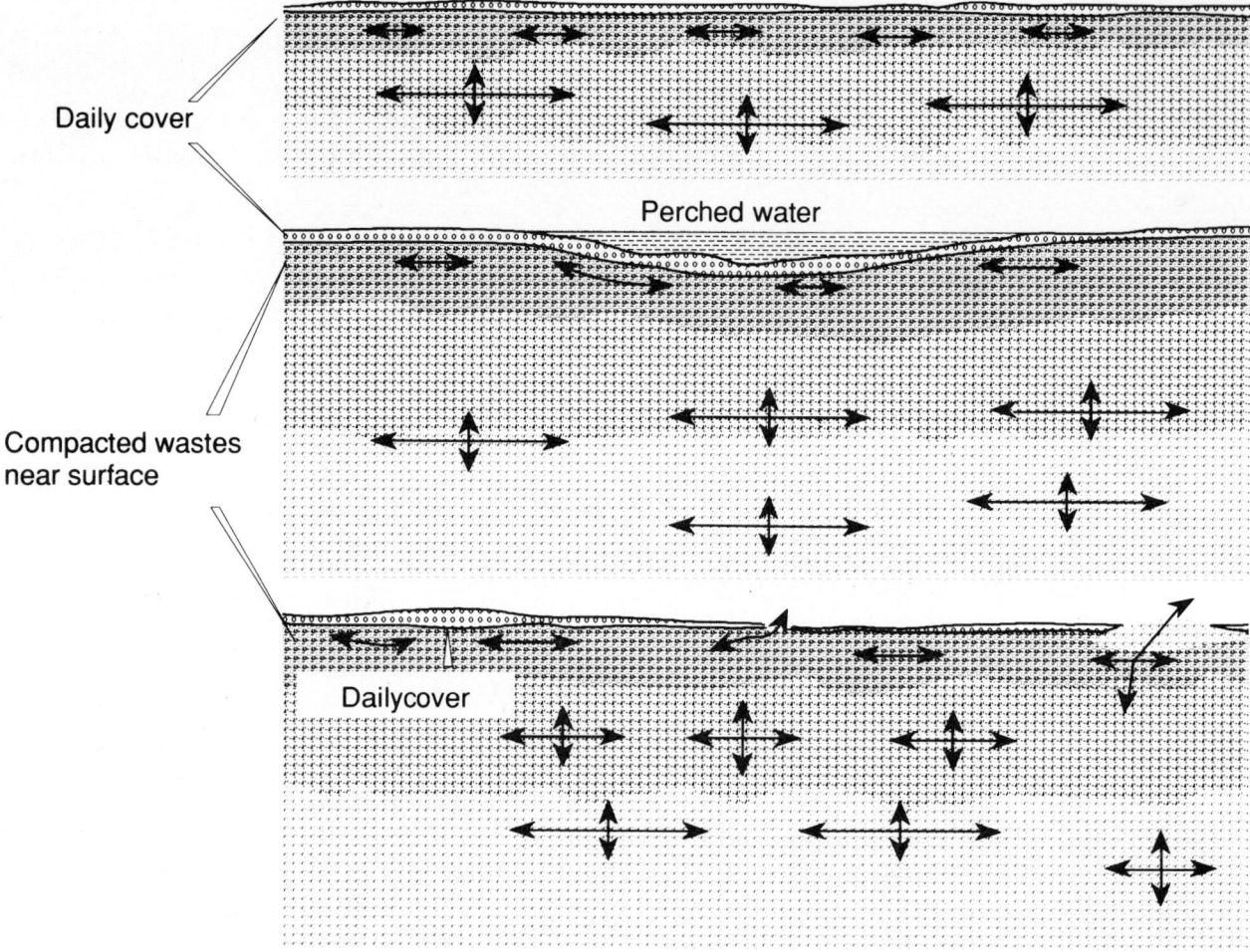

Figure 5.1 Gas movement within wastes

gas pressure building up, leading to uncontrolled migration of gas from the site. Gas vents constructed through the cap could, if not properly designed, allow water to enter the site. Climatic conditions may also encourage gas migration through the cap unless appropriate measures are taken. For example during very dry weather a clay cap may become cracked, especially at site boundaries, allowing gas to escape to the atmosphere.

5.5   Some landfill sites in disused quarries may have been partly backfilled with mine and quarry wastes prior to the disposal of controlled wastes. Unless the nature of these materials is known, and how they were deposited (eg loose tipped or layered and compacted) their suitability cannot be accepted either for forming an adequate basal seal for a subsequent landfill or for providing a safe route for passive venting of gases to the surface.

Migration outside the site

5.6   The potential for the migration of gas beyond the site boundary will depend on the geological characteristics of the adjoining strata, coupled with any man-made pathways such as mine-workings, service ducts, drains and blasting fractures. Natural pathways include permeable rock (consolidated or unconsolidated), planar openings such as joints, faults, bedding planes and cavities such as limestone cave systems. Figure 5.2 illustrates some gas migration pathways.

5.7 In varied lithologies gas will tend to migrate preferentially through beds of rock whose grain size, shape and packing are such as to make them most permeable. Gas may reach the surface some distance from the site by travelling through these strata and could then pass into service ducts or buildings. At sites where no control measures have been installed landfill gas has migrated 300 to 400 metres outside the site, and on some occasions, further. The permeabilities of strata are rarely consistent and even very low permeability clays may contain zones of higher permeability such as sand lenses. The per-

Figure 5.2 Possible gas migration paths from a completed/restored site

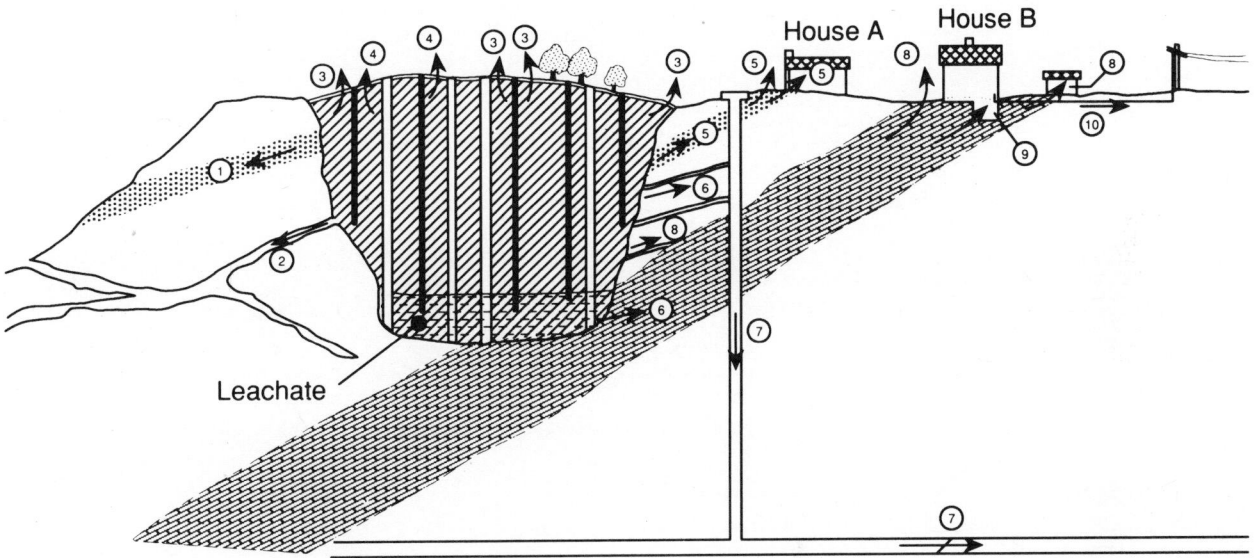

Gas pathways to atmosphere
① Through high permeability strata down the bedding plane

② Through caves/cavities

③ Through dessication cracks of the capping at the site perimeter, around tree roots, etc

④ Around site features which provide vertical pathways; gas or leachate wells

⑤ Through high permeability strata up the bedding plane, to atmosphere or house A

⑥ Through fissures caused by explosives etc

⑦ Along man made shafts etc

⑧ Through highly fissured strata into the atmosphere or buildings such as house B or shed etc

⑨ Into underground rooms

⑩ Along underground services

Notes
i) Gas may vary depending on its source from within the landfill and the migration route, eg route ⑤ gas compared to route ⑧ gas

ii) Leachate may degrade to give rise to gas generation at some distance from the site meability to gas of a given stratum is considerably greater than its corresponding permeability to water. The degree of saturation of the strata affects the permeability, for example dry clays are very permeable compared to wet clays.

5.8 At sites with an unsaturated zone beneath the base of the site there is the possibility that gas will move down into this zone before lateral migration takes place. This is an important consideration for the deposit of biodegradable wastes above ground level, and for "attenuate and disperse" sites. There is also a possibility that leachate could generate or evolve dissolved gas after migrating from these types of sites. The different solubilities of the component gases may also affect the composition of the gas from this source. In all such cases an assessment of the risk is necessary.

5.9 Rocks contain faults, fractures and fissures through which gas can migrate as indicated in Figure 5.2. Blasting fractures often occur in the walls of hard rock quarries and can extend for many metres into the surrounding rock. The effect of fractures on gas migration and the possibility of sealing them should be assessed as part of the site selection process. Some rocks, particularly limestones, contain cavities where potholes and cave systems can extend for miles. It is essential that all identifiable cavities in hydraulic continuity with the landfill that can be identified are sealed prior to deposit of wastes, not only for the control of gas but also of leachate.

5.10 Landfill sites are often located in areas where coal and other minerals have been mined and where ancient methane gas of geological origin may be present. These areas often contain underground mine-workings where disused tunnels and shafts can act as conduits for gas. Proposed sites should be engineered to prevent landfill gas migrating into mine-workings and vice versa, and additional sampling and analysis will be necessary to distinguish the different sources of gas (see Paragraphs 6.3–6.5 and 6.9).

5.11 Other major man-made pathways for gas migration from a site are underground service ducts used for electricity, telephone, TV cables, street lighting cables, water and gas pipes, sewers, drains, and land drains. At operational sites special attention needs to be given to migration along site roads and into weighbridge pits. The location of statutory services may be obtained from the relevant authorities. Land drains which are often installed in areas of poor natural drainage can be difficult to locate, although in agricultural areas it may be possible to locate them by aerial photography.

Gas accumulations

5.12 As well as providing pathways for landfill gas migration, several of the features described above can lead to voids where gas can accumulate. These include:– buildings and structures with service inlet ducts; wall and underfloor cavities; cellars and basements; cable chambers in telephone exchanges or where electric switchgear is located; the space under site weighbridges; lighting columns; drains, soakaways, stone infills, manholes, sewers or other pipes; settlement cavities beneath foundation rafts; behind skirting boards, cupboards or built-in units.

Changing patterns of migration

5.13 Changes in the migration pathways can occur over time due to changes in the accessibility of the pathway or in the evolution rate of the gas. Pathways change with time due mainly to physical changes such as settlement or surface sealing; eg, by ice, snow or by fine particles carried by water run-off. The pathways can also be affected by changes in permeability of surface materials during capping, restoration or soil compaction and by changes in the water table or leachate levels (especially after cessation of pumping leachate or the extraction of adjacent groundwater). The extension or collapse of mine-workings, or natural cavities, can also change pathways.

5.14 Gas composition may be changed by reaction along the gas migration pathway. More complex trace components may be removed by reaction after a short distance along the migration pathway. Carbon dioxide can be taken up by adsorption, absorption or dissolution, thereby increasing the methane concentration of the gas. Alternatively, microbial oxidation of the methane can occur which increases the concentration of carbon dioxide. Such oxidation is exothermic and may be associated with "hotspots" in the affected ground. When changes in pathways occur there is a possibility that, due to these reactions, gas composition may alter.

5.15 Changes in atmospheric pressure can increase or decrease volumes of gas migrating from a site, and therefore need to be considered in relation to gas monitoring. A sudden fall in atmospheric pressure will increase the pressure differential between the waste and the atmosphere, tending to draw gas from the ground. Conversely, gas release may be reduced during sudden rises of atmospheric pressure. During sustained periods of steady atmospheric pressure, and if other factors are constant, a fairly regular flow will occur. Gas pressures within wastes will be affected by rapid changes in leachate levels. These effects need to be evaluated in relation to the relative permeabilities of the landfilled wastes, the final cover and any lining materials. They should also be compared with the permeability of the surrounding strata. Where possible, quantitative data on the permeability of waste, cover materials, linings and surrounding strata should be included in the assessment of the site. The differential between atmospheric pressure and gas pressure in the surrounding strata may be more influential on gas migration potential than the pressure differential between the landfill and the air mass above it.

CHAPTER 6

Site assessment

Introduction

6.1 In order to design effective gas control and monitoring systems, information is required on many aspects. A survey of the site should seek to identify the sources, volumes, composition and evolution rates of gas; identify likely migration routes; and assess risk to surrounding development, crops and vegetation. The planning of this survey requires great care and specialist advice may need to be sought. Such information will also be necessary in the event of an Environmental Assessment being required under the relevant planning regulations (see Appendix B).

Closed sites

6.2 Owners of land containing restored sites, and sites awaiting restoration, are advised to investigate for the presence of landfill gas on their property and notify their local WRA of the former site. Specialist advice may be required to assess whether the wastes are evolving, or likely to evolve, gas in quantities and concentrations that may present a hazard.

6.3 A desk study should be undertaken to examine the geology and hydrogeology of the site and its environs and to evaluate the effect these may have on gas migration. Topographical information should also be obtained, particularly the position of development, underground services and public rights of way. Where buildings are considered to be at risk, the methods of construction, provision and location of services should be examined. In mining areas, contact should be made with the mining company, such as British Coal, to obtain details of strata, underground workings and whether the landfill site has been used for the disposal of spoil. Where landfill sites are contained within old quarry workings, then all available information should be gathered to identify areas within the quarry which were backfilled with quarry spoil. As much detail as possible should be obtained of the history of the site, types of wastes deposited, locations of any hazardous waste, the methods of operation, and any pollution control measures provided. Desk studies are especially important in deciding priorities, when a number of sites need to be examined within the resources available for the work.

6.4 The relevant local authorities for any area are sources of advice on planning matters and development proposals. Many authorities which have extensive mined ground in their areas maintain records of the extent of mined ground and the locations of mine openings. British Coal Corporation has detailed records of those coal mines for which they have a direct or an inherited responsibility. Records in respect of the "free" mines for coal in the Forest of Dean are held by the Forestry Commission at Coleford, Gloucestershire. Important sources of information and advice on mined ground in general are the Health and Safety Executive and the British Geological Survey.

6.5 A preliminary physical appraisal of the site and its environs should be undertaken. This should seek to check and add to the information obtained in the desk study. A vegetation survey may assist by the identification of characteristic areas of die-back. A surface investigation should be conducted for the presence of flammable gas both over and around the site. Subsurface measurements should then be undertaken by examination of atmospheres within manholes (see para 7.29), ducts etc, or by subsurface probes. This search should be carried out by sampling all the gas migration pathways and locations where gas may accumulate, identified in the desk study. Where no evidence of gas is found it cannot be concluded that no gas is being evolved.

6.6 Where there is no development proposed or at risk from a site and no evidence of gas generation or of gas likely to be evolved, then the assessment may be concluded.

6.7 At existing or proposed sites where development surrounds or is on the landfill, or where the desk study indicates that the landfill is likely to produce large volumes of gas, then an appropriate geological and hydrogeological investigation of the site and its surroundings should be undertaken. Much of the design of the survey should follow the advice contained in Paragraphs 3.109 to 3.119 of Waste Management Paper No 26 but extended to a distance of at least 250 metres from the site boundary (see also the statutory requirement to consult Waste Disposal Authorities under Article 18(i)(w) of the Town & Country Planning General Development Order 1988—not applicable to Scotland). The survey will need to be extended beyond 250 metres if the geology or man-made features are likely to provide migration pathways beyond this limit.

6.8 For sites with some development within 250 metres of the landfill, a survey should be undertaken of the landfill and also the area between the landfill and the development.

6.9 It is unlikely that a survey would be adequate unless boreholes and wells are drilled both within and outside the wastes to provide geological and hydrogeological information on the environs of the site and details of waste type, leachate depths, gas quality and quantity and temperature within the landfill. In fractured or fissured strata, no guarantee can be given that all fissures/fractures are intersected by boreholes. It is essential therefore that an adequate number are drilled to provide an assurance that a representative proportion of migration pathways are monitored so that an assessment can be made of risk from escaping gas. Specialist advice should be sought on borehole spacings, sampling frequency and interpretation of results.

6.10 For gassing sites near to development, control measures may need to be introduced as a matter of urgency. It is also essential that all underground services near to or on the site are regularly checked for the presence of gas and all relevant undertakings are informed of the risk. Such undertakings are advised that when working on underground services near landfill sites, precautions are taken by employees to minimise any risk from the presence of landfill gas.

6.11 Whatever the findings of the site assessment, landowners are recommended to keep a copy of the results of the desk study and site appraisal.

6.12 Where development is proposed and where there is no information available on landfill gas evolution, an investigation should be undertaken as detailed in Department of the Environment Circular 21/87 (22/87 in Wales and Planning Advice Note 33 in Scotland). If landfill gas is found then specialist advice should be sought to assess whether it is likely to pose a risk to public health.

Operational sites

6.13 Operators of operational sites should follow a similar course of action to that described in Paragraphs 6.3 to 6.9, in consultation with the WRA. Any monitoring and any control measures considered necessary should be included in the working plan and licence conditions for the site.

Obtaining a database

6.14 When assessing gas evolution and emission from a landfill, it is important that measurements are repeated to establish an adequate database for interpretation. Due to the variations that occur in gas composition, a single set of measurements is not adequate. A series of measurements is necessary over a period of time. Measurements should be taken in different weather conditions, especially in relation to atmospheric pressure. A pattern of gas concentrations in and around the site should be established. After gas control measures have been instituted, these measurements should be repeated and the results cross referenced with the database so that any anomalies can be investigated. While gas composition will be a primary requirement of the database, additional information such as pressure and flow rate are important together with meteorological conditions, especially when designing a control or gas utilization scheme. A knowledge of gas temperature may also be useful. For operational sites, the level of monitoring required will increase as waste disposal progresses, to establish an adequate database.

Proposed sites

6.15 The geological formations surrounding a landfill will have a significant influence on the design of gas monitoring strategy as well as the necessary control systems. Accordingly no site should be selected, and certainly not authorised by licensing, until the assessment of the site and its surroundings, as described in Paragraphs 6.3 to 6.9 above, has been undertaken. Although it is possible to engineer many geologically sensitive sites to a standard acceptable for proper landfilling, this is not so for all potential sites.

6.16 It is essential that comprehensive borehole and surface monitoring is undertaken prior to the deposit of any wastes. This will provide information on natural sources of gas around and under the site. As a guide, a minimum of twelve data sets should be collected within a minimum of a three month period to give the necessary understanding of background concentrations.

Other sources of gas

6.17 In certain areas near coal measures, high concentrations of methane have been detected in boreholes before wastes have been deposited. There have also been examples of high concentrations of naturally occurring carbon dioxide. It is therefore essential to determine the sources of gas if a suitable gas control system is to be designed and effectively monitored. Where high concentrations of methane or carbon dioxide are found to arise from such external sources, the "base line" for detection of migrating gases from a site will need to be varied accordingly.

6.18 Coal mines and coal measures, marshes and natural gas pipelines are all possible alternative sources of methane in addition to landfill gas. Where one or more of these is present close to the landfill, if flammable gas is found, analysis by gas chromatography usually can be used to establish the source of the gas. This can be achieved by examining for the presence of trace gases to fingerprint the source of gas detected in an investigation. Propane and helium are examples of gases which are indicative of alternative sources of methane to landfill gas. Information on alternative sources of gas is provided in Appendix VIII of Annex 2 of Waste Management Paper No 26. When gas chromatography fails to establish the source of flammable gas then carbon dating may establish its origin, provided the gas sample is not a mixture from different sources (See Appendix F).

6.19 High concentrations of carbon dioxide occur naturally at shallow depths of up to 2 metres due to microbial activity associated with the roots of many types of vegetation, providing concentrations of up to 7% by volume in certain soils (silty clays). At greater depths high concentrations of carbon dioxide may arise from the action of acidic water on limestone rocks or due to microbial activity above certain sulphide containing mineral veins. There may therefore be a wide range of background concentrations of carbon dioxide in the ground around sites which need to be considered.

6.20 Where it is thought that gas other than landfill gas is present in samples, near to a site where wastes have been deposited, an investigation should be undertaken to determine the background concentration in the locality. This can usually be achieved by the installation of monitoring points near the site where there are similar soils and geology etc., but in locations where it is unlikely that there are migration pathways from the landfill.

CHAPTER 7
Monitoring for landfill gas

Introduction

7.1 A site specific monitoring programme should be established at every site where controlled waste has been, is being or is to be deposited. The aim is to provide information to assess whether gas generation from the wastes is likely to give rise to risk to public health or the environment, or a nuisance. Where control measures are installed at a site, monitoring should seek to check that the measures continue to remain effective and to measure any loss of efficiency in the control system. For sites without control measures, monitoring should be targeted at measuring for changes in gas evolution and migration patterns. When monitoring establishes that adverse changes are taking place at a site, then further assessment is required to determine the action to be taken. Monitoring will be needed at most sites for a number of years after the site has been closed and restored. It may be possible to stop when gas concentrations have declined below the levels specified in Paragraph 7.9. If any significant change occurs at or near a site (building developments, change in water table levels etc), which may affect gas evolution or migration patterns, then further assessment including monitoring will be required. Further information on gas monitoring can be obtained from the Institute of Wastes Management publication "Monitoring of Landfill Gas" and from "Measurement of Gas Emissions from Contaminated Land" by the Building Research Establishment.

7.2 A gas monitoring programme should be incorporated into the design of all operating and proposed landfills irrespective of waste type. This will include the provision of monitoring points needed beyond the site boundary and particularly between the landfill and any nearby development. The location of monitoring points will be determined from the site survey.

7.3 The safety of personnel involved in monitoring will require special attention and it is recommended that safe working procedures be instituted by all operators and WRAs. All employees involved should be instructed in the detail of these procedures. These should include consideration of the control of exposure of operators, by inhalation, to landfill gas issuing from boreholes and wells, taking into account the characteristics of the gas described in Chapter 3. In particular, entry into any space where there is a risk of poor ventilation should be prohibited unless the precautions detailed in the Health and Safety Executive Guidance Note GS5 entitled "Entry into Confined Spaces" have been followed.

Monitoring frequency

7.4 The frequency of monitoring required will be site specific, and will depend on a number of factors. These include:–

(a)   The age of the site;

(b)   The type and mix of waste;

(c)   The possible hazard or nuisance from gas escaping from the site;

(d)   The results of previous monitoring;

(e)   The control measures that have been or are to be installed;

(f) The development surrounding the site; and

(g) The geology.

It has been found that many sites which have taken supposedly only "inert" wastes have evolved gas. It is therefore important not to rely on information about waste type unless there has been exceptionally good control of waste input and comprehensive records exist.

Closed sites

7.5 The majority of closed sites are unlikely to have monitoring boreholes or wells. Where it is concluded by the site assessment that no gas is being or is likely to be produced and there is no risk to development, then there should be no need to undertake monitoring (see paragraph 6.6). For other sites the frequency of monitoring should be determined by the site assessment. As a guide, it is recommended that rural sites with no control system installed should be monitored four times per annum. This frequency should be increased at sites with control systems. Monthly inspection is recommended where there is building development or there are services within 250 metres of the deposited wastes. Sites should be monitored at least weekly where any development is within 50 metres of a waste deposit containing significant amounts of biodegradable materials. Where gas migration has been identified, more frequent monitoring may be necessary.

Operational sites

7.6 The site assessment should determine the frequency of monitoring for sites. As a guide however, for isolated rural sites, where quantities of biodegradable wastes have been deposited, gas monitoring frequency should, initially, be on a monthly basis. For routine monitoring this may be extended to three monthly intervals. For all sites taking or that have taken biodegradable wastes that have buildings or services within 250 metres of the limit of filling, monitoring should initially take place at least weekly or more frequently if gas migration has been identified. The monitoring frequency should be subject to regular review.

7.7 At all sites, the risks to persons involved in site operations should be specifically considered by the site operator, and monitoring of on-site buildings, enclosures, confined spaces, cavities, man-holes etc, should be carried out when and where necessary. Such monitoring may need to be very frequent, possibly on a daily or continuous basis.

7.8 Site monitoring frequencies should be varied under certain conditions. They should be increased when:–

a) increases in gas quantity or changes in gas quality are found during routine off-site monitoring;

b) control systems are changed by landfill operations;

c) capping of part or all of the site takes place;

d) pumping of leachate ceases or starts or leachate levels rise within the wastes;

e) variations in weather occur when gas migration pathways can be changed significantly (eg when ground freezing is widespread); and

f) building development on or adjacent to the site takes place. Additional monitoring points may also be necessary if this occurs.

Where regular monitoring has shown that conditions at a site are regular and a predictable pattern occurs, then the frequency of monitoring may be reduced. However, it is recommended that the interval should never exceed six months.

7.9 Monitoring should continue ideally until the whole of the biodegradable substrate within the wastes has been consumed. However, this could only be determined by examination of all the waste within the landfill. Monitoring should therefore continue until:–

a) the maximum concentration of flammable gas from biodegradation within the landfill remains less than 1% by volume (20% LEL) and the concentration of carbon dioxide from biodegradation within the landfill remains less than 1.5% by volume measured in any monitoring point within the wastes over a 24 month period taken on at least four separate occasions, including two occasions when atmospheric pressure was falling and was below 1,000 mb; or

b) an examination of the waste using an appropriate statistical sampling method provides a 95% level of confidence that the biodegradable matter has been used up.

7.10 Residual gas is often trapped within wastes in old landfills. Detection of methane or carbon dioxide may not necessarily indicate continuing gas generation. Therefore gas pressure or emission rates should be monitored and pumping carried out to determine whether the gas source becomes depleted. Monitoring can then be undertaken to establish whether gas recovery occurs.

7.11 Where development is proposed on or within 50 metres of sites and consideration is being given to cessation of monitoring, then sample drilling may be appropriate to provide additional information on the condition of the wastes.

7.12 At sites where analyses of core material recovered from the wastes have revealed that none of the wastes are biodegradable, but very low concentrations of landfill gas constituents are found, then monitoring may be stopped if the relevant Authorities (Planning or Waste Regulation) are satisfied that there is no significant risk.

Weather services for landfill gas management

7.13 The Meteorological Office can advise on all aspects of the weather. Forecasting services are available from 14 Weather Centres across the UK. These services include:–

a) Warnings of atmospheric pressure falls. These are issued when the pressure is expected to drop at a specified rate, (eg 4 mb or more in 3 hours);

b) Forecasts of atmospheric pressure. Forecast of pressures up to 36 hours ahead, giving a trend beyond if required;

c) Forecasts of heavy rainfall, snow or severe frost. For 24 hours or more ahead, the precipitation forecasts are indicative rather than precise for a specific site, but at short range, precision improves (using information from the rainfall radar network).

Past weather information is collected from a large number of stations located across the country. This information is available from Met Offices at Bracknell, Edinburgh and Belfast and can be provided on the following subjects of interest:–

d) Hourly atmospheric pressure may be summarised (eg in terms of pressure changes);

e) Rainfall amounts—hourly/daily/monthly;

f) Soil moisture deficit and evaporation—daily/weekly/monthly.

Supervision of monitoring

7.14 All monitoring should be undertaken and supervised by staff who are knowledgeable and experienced in the use of equipment and interpretation of data obtained. They should know what monitoring equipment is available, how to use it, and be aware of its limitations.

Responsibility for monitoring

7.15 For closed sites, the responsibility for assessment and monitoring of the site will rest primarily with the landowner. The Environmental Health Authority should attempt to identify such sites and advise the landowner to undertake an assessment of the site with respect to gas evolution and migration. The Environmental Health Authority may also wish to investigate these sites. The WRA should be consulted where technical advice is required.

7.16 Section 61 of the Environmental Protection Act 1990 provides a duty on the WRA to arrange for regular inspection of its area to detect whether landfill gas is causing pollution of the environment or harm to human health. Where a risk is identified, the WRA is required to take the necessary steps to avoid such harm or pollution. When the relevant sections of the Environmental Protection Act 1990 are in force the responsibility for monitoring all operational and proposed sites under a waste management licence will rest with the licence holder until a "Certificate of Completion" has been issued by the WRA. The WRA must confirm that monitoring is taking place and carry out its own measurements at regular intervals. Existing licences should require the establishment of an adequate monitoring scheme. Monitoring for landfill gas should be continued until the certificate of completion has been issued.

7.17 Where any changes are proposed at a landfill site involving the need for planning permission, then an agreement as provided for by the Town and Country Planning Acts should be considered. The agreement should ensure adequate monitoring and control of landfill gas until gas evolution has ceased, as defined in Paragraph 7.9. In such circumstances, it will be the responsibility of the Planning Authority to ensure that monitoring and control is taking place. Responsibility for confirming and assessing the results of post-closure monitoring rests with the local Environmental Health Authority.

7.18 When development takes place on or adjoins a landfill site, responsibility for the monitoring in respect of the safety of employees involved in the development and the occupants of the property falls to the developer, who should also take into account the advice given in the papers of the Inter-Departmental Committee on the Redevelopment of Contaminated Land and the approved building codes. For further discussion on this topic see Chapter 9.

Review of the monitoring strategy

7.19 The monitoring strategy should be kept under regular review and subject to specialist appraisal. The intervals of such assessment will be site specific but should certainly not exceed 12 months.

Emergency monitoring

7.20 It will occasionally be necessary for the operator, Waste Regulation and Environmental Health authorities to undertake monitoring as an emergency response where gas is suspected, or known, to be migrating from a landfill site.

There is also a possibility that routine monitoring might identify dangerous concentrations of gas in or adjacent to development. The procedure given in Appendix C for monitoring gas in buildings should be followed. The immediate response should be to assess the safety of all occupants of the property. Action should be taken to identify the source of the gas and the point or points of gas ingress into the premises. When these have been established, then appropriate action can be taken although the effectiveness of this action should be confirmed by monitoring. If any concentration of flammable gas is found in a building which may be attributable to a gas leak from piped gas, then regulations require that the local public gas supplier (usually British Gas plc) should be informed immediately (Gas Safety (Installation and Use) Regulations 1984).

Trigger values in buildings

7.21 It is difficult to define precisely those concentrations of landfill gas which should be considered dangerous when they occur in buildings. Risk depends not only on the concentrations of gas in the building, but also the volume of space affected, the occupancy and the control of access to the building, and the ease with which the accumulation of gas can be immediately dispersed by the provision of appropriate ventilation. Furthermore, the existence of enclosed or poorly-ventilated void spaces which may be affected by gas may not be known or apparent. Buildings should accordingly be evacuated when concentrations of methane or flammable gas above 1% v/v and/or 20% of the L.E.L., are found in any void space, eg. rooms, cellars, basements, integral garages, roof cavities, cupboards, underfloor cavities etc. which are occupied or are near to occupied areas. All people should leave any area where concentrations of carbon dioxide exceed 1.5% by volume. When such concentrations are found, HSE, emergency services, the local Environmental Health Department and the WRA should be informed. Appropriate steps should be taken to implement precautions and control measures. Action to be taken is described in Appendix C.

Potential for fires on landfill sites

7.22 There is generally little risk of an explosion from landfill gas venting to atmosphere, though there may be a possibility of a flash fire if accumulation occurs in poorly ventilated areas. This risk can be minimised by allowing gas to vent only from controlled points. The surface of a restored gassing landfill should be monitored regularly to confirm the continued integrity of the cap. Where high gas concentrations are found in cracks and fissures, access to the area should be prevented by fencing or alternative measures should be taken, such as repair of the cap etc. High concentrations of gas are likely to be found on the surface of unfinished sites and operators need to take precautions accordingly (see Paras 8.45 to 8.51).

Portable instruments

7.23 Portable instruments are best used in reconnaissance or preliminary surveys and to provide an indication of likely gas concentrations. There is a variety of portable instruments available which are described in Appendix D. Their limitations should be noted and care taken in their selection and use especially with respect to the reporting of the appropriate values measured which may be methane/flammable gas or % LEL according to the circumstances and equipment used. Particular attention should be paid to regular calibration and maintenance. The accuracy of field readings should be confirmed periodically by taking duplicate samples for laboratory analysis by gas chromatography (GC). GC or infra-red methods are currently the most suitable for obtaining precise analysis of the gas sample. The accuracy of contract analysis should also be checked occasionally by use of a reputable independent laboratory. Those laboratory facilities accredited under the NAMAS scheme (The National Measurement Accreditation Service) or that can demonstrate rigorous Quality Assurance procedures are most likely to provide accurate results.

Gas sample collection

7.24 Great care must be exercised when sampling and particular attention should be paid to the method of sample collection, which is a crucial element in determining the performance of a gas monitoring scheme. Sample collection methods, including the types of containers to be used, are described in Appendix E.

Other aids to monitoring

7.25 There are several other aids to monitoring which could be used on specific occasions. These include carbon dating, measurement of other parameters relating to the gas and some aerial techniques. These are described briefly in Appendix F.

Monitoring techniques

7.26 Several techniques may be used for monitoring landfill gas. These include:–

a) Surface Monitoring;

b) Subsurface monitoring (Gas probes);

c) Excavated pits and trenches backfilled around standpipes;

d) Gas monitoring boreholes or wells; and

e) The use of leachate wells.

7.27 Monitoring using specially constructed wells and boreholes is the preferred method. At completed and shallow sites adequate monitoring in the short term may be achieved using excavated pits or trenches. At every site where there is a potential for gas to migrate to development or underground services, monitoring boreholes should be installed at appropriate points outside the wastes and between the filled area and any development which is potentially at risk.

Surface monitoring

7.28 Surface monitoring is carried out with portable instruments to assist in determining the likely presence of gas escapes. Confirmation of the source of the gas will then be required, usually by GC analysis, for which samples should be taken. Surface monitoring should be used to: check the integrity of caps on the waste; check on borehole monitoring; aid the siting of monitoring points; monitor for the presence of gas during filling and give preliminary indications of off-site migration.

Sub-surface monitoring—gas probes

7.29 The position of underground cables or gas pipes should be determined from plans and by use of cable location devices before probes are inserted. To ensure the safety of persons involved in this work, the guidance provided by the HSE booklet number HS(G)47 entitled "Avoiding danger from underground services" should be followed.

7.30 The use of probes driven into waste or strata enables point source monitoring of gas concentrations in the local environment around the probe to be carried out. The design of such a probe is illustrated in Figure 7.1. Various probes are available, consisting usually of metal tapered tips coupled firstly to short perforated pipe sections, and then to longer unperforated metal pipes. They may be driven into soils or wastes usually down to depths of about one to two metres, although it is possible to achieve depths close to 10 metres with specialised equipment. It should also be recognized that gas volumes collected within the probes are small, and care should be exercised in extracting samples for analysis to ensure that air is not introduced. Most monitoring devices require a flow of gas past the detector or sample collection device, which is often not available from this type of probe.

7.31 Other methods are often adopted for obtaining subsurface samples such as the use of an insulated searcher bar driven into the ground by a sledgehammer and a mild steel driving cap. It is withdrawn and a plastic tube approximately 1.5 metres in length (usually made from 12 mm outer diameter PVC tubing with the lower metre perforated with 6 mm holes, at 100 mm intervals, drilled in a spiral) is inserted into the hole. The ground is normally sealed using the sledgehammer, and a cap fitted over the end of the tubing (see Figure 7.2). Augers can also be used to provide a similar sampling point. These probes suffer similar limitations to those described above.

7.32 The use of these types of sampling probes cannot ensure that penetration into the fill is achieved, as the depth of the cap may vary considerably. The operator should determine and record the sampling horizon. The permeability of soil and cap materials may be so low as to prevent an adequate flow of gas to the measuring device and may not be representative of the soil gas environment. Similar difficulties may be encountered where the water table is close to the ground level.

Sub-surface monitoring—excavated pits and trenches

7.33 Excavated pits and trenches provide a means for monitoring gases in shallow sites. Perforated or slotted plastic tubes (usually around 50 to 80 mm in diameter) are placed within the pit or trench and surrounded with a granular medium. The excavation is then backfilled and the surface is sealed eg with a bentonite or clay cover. As air will inevitably be introduced to the waste during construction, monitoring should take place over a number of weeks, especially where gas evolution rates are low.

Sub-surface monitoring—gas monitoring boreholes and wells

7.34 The preferred method for landfill gas monitoring is by using properly designed and constructed boreholes or wells that are dedicated to this function. During active gas abstraction, wells should not be used for leachate monitoring without safeguards to prevent air being drawn into the gas abstraction system. Gas monitoring wells should be installed within landfilled wastes and boreholes outside the fill area of landfills, where a potential risk has been identified.

7.35 A systematic method should be used for monitoring and sampling at boreholes and wells to ensure consistency of results. Appendix G provides an example of a procedure for monitoring at boreholes and wells in and around sites.

7.36 In their simplest form boreholes and wells consist of perforated plastic casing abutting directly to the strata (Figure 7.3). Probes or tubes may be permanently installed (or inserted at specific depths) within the casing. Depending on the amount and concentration of gas in the vicinity, such sampling may only give indicative values for average concentrations throughout the depth profile of the borehole or well. They should be capped to prevent air ingress and protected against vandalism or accidental damage.

7.37 More representative data may be obtained by installing discrete sampling probes at set intervals within boreholes or wells (Figure 7.4).

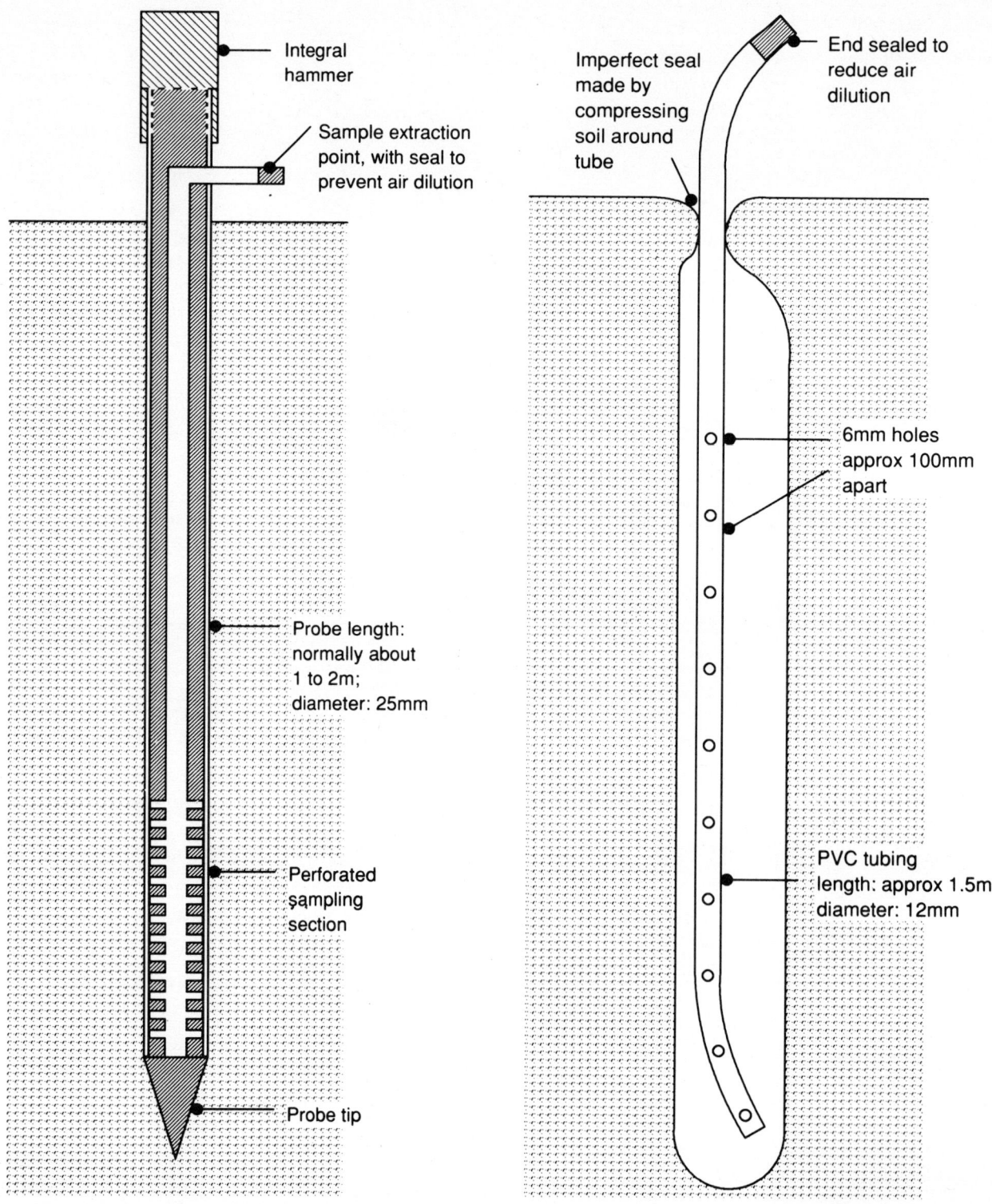

Figure 7.1 Sub-surface probe (steel pipe)   *Figure 7.2 Sub-surface probe (plastic pipe)*

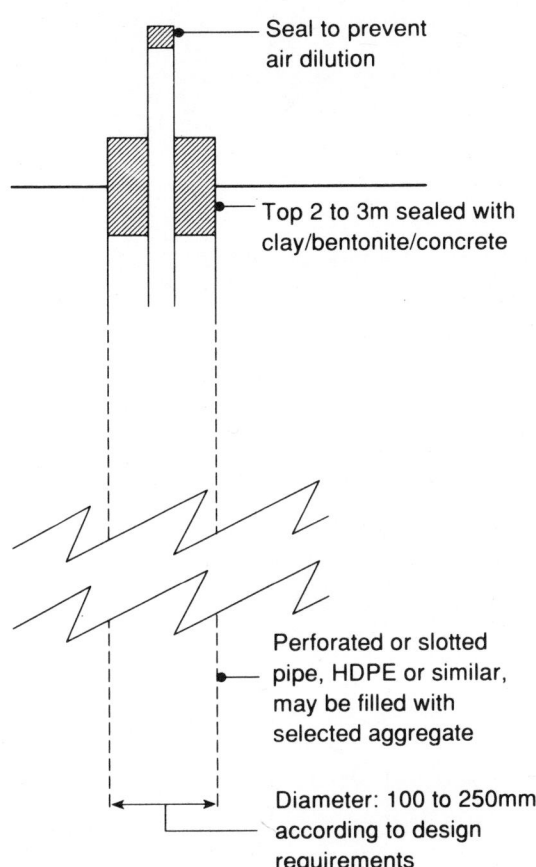

Figure 7.3 Simple monitoring borehole

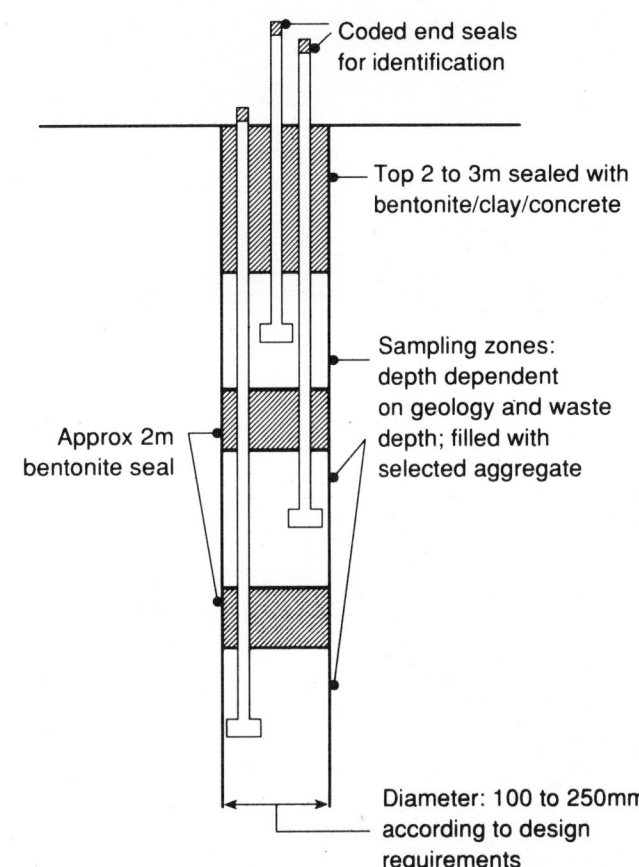

Figure 7.4 Multiple point monitoring borehole

The probes are surrounded by carbonate-free gravel of uniform size and isolated from each other by a column of impermeable material, at least 2 metres thick. Sampling from probes at locations outside the site can provide data on migration potential at the sampling depth. Information on gas evolution (at the sampling depth) is obtained from wells drilled within the wastes. Care should be taken not to extract gas at too high a rate to reduce inaccuracies in sampling. Alternatively, separate adjacent boreholes or wells may be used, each drilled to a different depth. This may be necessary where horizons have varying degrees of gas permeability and where it is essential to identify concentrations in specific rock layers.

7.38 Where wells are drilled into landfilled wastes within sites, there are several additional points to be borne in mind. Evidence of water levels encountered during drilling need not be indicative of ground water/leachate levels. A perched water table may exist above low permeability material such as a clay layer or compacted daily cover, and permeable wastes may still exist below. Thus all in-site monitoring wells and gas abstraction wells should be installed close to the base of the deposited wastes, taking care not to breach any containment provided on the base. Exceptions to this will apply where water/leachate levels are known, and where the leachate level is likely to remain constant. Wells drilled within sites can provide information on gas composition, temperature, pressure, flow rates and waste composition, but will not provide any evidence of lateral migration. The measurement of flow rates is not an easy task. Information on this is given in Waste Management Paper No 26 (Annex 2 paras. 7.8-7.10). Unless the wells contain multilevel sealed probes, they can only provide indications of average conditions throughout the waste profile.

7.39 When drilling through perched water tables in wastes, care should be taken that the well does not become flooded. The use of a casing during drilling can prevent the flow of liquid into the borehole. Examination of the extracted waste may enable the degree of saturation of the waste beneath the perched water table to be determined.

7.40 Any investigation involving the disturbance of ground in which landfill gas is suspected to be present must be carried out with appropriate safety precautions. A written safe system of work with rehearsed emergency procedures should be provided before work starts and it should be rigorously followed.

7.41 Gas concentrations should be monitored during drilling operations. Electrical equipment should be sited in a safe area, preferably at least 4 metres from the top of the well. Where it is necessary to locate electrical equipment in a hazardous area, it should be suitable for use in flammable atmospheres, and conform with the requirements of BS 5345. Other sources of ignition eg. naked lights, should be removed from the vicinity of drilling operations, and no smoking signs posted in the area. Operators should be trained in emergency procedures.

7.42 Instructions should be issued on what action to take in the event of dangerous substances being encountered during drilling. Drilling operations should avoid, as far as possible, any area where dangerous substances are known to have been deposited. However this will not prevent the occasional "rogue" deposit being brought to the surface. All operators should be issued with, and keep available for use, suitable personal protection equipment, which they have been trained to use.

7.43 There are no substantial differences between the design of monitoring wells to be used inside or boreholes outside the site, although wells constructed within sites may need to be of a larger diameter to compensate for difficulties encountered in penetrating the infilled wastes. Whilst wells within wastes should not penetrate the base of the site (see Para 7.38), an appropriate number of the boreholes drilled outside the wastes should be drilled to a depth below the waste deposits, depending on site geology and hydrogeology.

7.44 Drilling wells or boreholes will inevitably disturb the ground, leading to temporary aeration. Accordingly sufficient time should be allowed for the system to equilibrate (two to three weeks is not uncommon) before sampling. Some augers have sample probes fitted which allow gas measurements to be taken as the drilling proceeds. Measurements taken in this manner should only be used as an indication of the presence of gas to assist in determining the strata that require monitoring.

7.45 Borehole location and spacing outside the wastes is site specific. Their spacing around sites is related to the likely risks posed by the gas being evolved. The risk will vary with the gas quality and volumes being evolved, the gas permeability of the wastes, any site engineering works (ie. surface capping and any site linings) and the surrounding strata as well as the proximity of buildings and services. The spacing of boreholes around the site will rarely be uniform, as more boreholes should be drilled nearest to development or where there are changes in the site geology. If landfill gas is found in monitoring boreholes at concentrations in excess of those indicated in Paragraph 2.4, then additional boreholes between the perimeter boreholes and any "at risk" development may be necessary. For sites with a gas pumping system, monitoring boreholes should be located in the adjoining strata, midway between the gas extraction points. Table 7.1 provides a general guide to spacings between boreholes around sites. When drawing up gas management schemes, it must be recognized that the table shows examples and the need for monitoring and hence borehole spacing, should be related to the risk assessed at the site.

7.46 The optimum pattern, design and distribution of gas monitoring boreholes and wells should be determined by specialist advice.

Use of leachate wells

7.47 Where leachate monitoring or abstraction wells exist within sites, these may be used for gas monitoring purposes, but cannot be regarded as being comparable with, or substitutes for, specially designed gas monitoring points. The safety of large diameter (0.5 metres) chambers must be ensured and maintained. Measurements taken in manholes may be affected by sampling error due to mixing with air. At depth the values obtained may be affected by partial or complete blockages of any perforations originally provided. Where these are covered by the leachate level the measurement may only be of the concentration of gas in the headspace.

7.48 Where new landfill sites are being developed, or proposals are under consideration, the design should include the installation of purpose-built gas monitoring boreholes. This will avoid any potential hazards arising from using leachate wells and will assist reliability of monitoring data.

Table 7.1 Guidance on Borehole Spacing

Site Description	Borehole Spacing	Comment
Uniform strata (no fissures): no development within 250 metres.	50 metres to no boreholes needed.	Dependent on quantity of gas generated and effect on vegetation.
Uniform strata (no fissures): development within 100 metres.	50 metres maximum nearest development.	Dependent on quantity of gas generated, risk to development and effect on vegetation.
Uniform strata: development adjoining site.	30 metres maximum nearest development.	Dependent on quantity of gas being generated, risk to development and type of strata.
Fissured strata: development within 250 metres.	20 metres to no boreholes.	Dependent on quantity of gas being generated and effect on vegetation.
Fissured strata: development within 100 metres.	Up to 20 metres nearest development.	Dependent on quantity of gas being generated, nature and extent of fissuring and risk to development and effect on vegetation.
Fissured strata: development adjoining site.	5 to 20 metres nearest development.	Dependent on quantity of gas being generated and nature and extent of fissuring.

CHAPTER 8
Control measures for landfill gas

Introduction

8.1 Gas should not be allowed to escape from a landfill in an unplanned or uncontrolled manner. It should be contained and preferably flared within the perimeter of the landfill site. It may be acceptable in certain circumstances to permit safe migration of the gas from the site or pipe it to a flare stack located in a secure compound off the site. Most sites should have an adequate gas management system installed, maintained and operated. Accordingly all relevant site licences should contain conditions relating to such systems. The need for them to be incorporated into planning consent conditions should also be considered.

8.2 A key requirement of a gas management system is that it should have adequate protection against failure. Its design is therefore a complex task requiring specialist advice. Each system will be site specific and accordingly only the main elements that might form part of such a system are described below. Single control measures are unlikely to be adequate except for some landfills in remote locations. At all other sites a combination of gas control measures will be required. Regular monitoring will also continue to be necessary for some time after the site has closed, possibly supported by remedial action.

8.3 The particular scheme that is adopted will depend on the expected pattern of gas evolution. For example, initial low volumes of gas may be passively vented, with active pumping and flaring or utilisation being employed later. In the final stages of landfill stabilisation, passive venting may again suffice. Thus the transition between these different control methods is a critical part of the design of the gas management system and should be managed accordingly.

Gas barriers

Operational and proposed sites

8.4 Clay or bentonite linings, synthetic membranes or grout curtains have been used as a barrier to restrict the migration of leachates from landfills. Waste Management Paper No 26 provides information on both natural linings and artificial membranes, the techniques of placing such material and their likely behaviour within the landfill. The effectiveness of liners alone in preventing gas migration is not yet fully quantified and their permeabilities to gas have yet to be determined. It would seem likely that they would be more permeable to gas by several orders of magnitude compared to leachate. Leachate containment systems should therefore never form the sole means of gas control at a landfill without a thorough assessment of their gas containment capabilities.

8.5 Reworked clay and certain bentonite linings are probably the most commonly available natural materials for gas barriers. They should be laid and compacted to achieve a coefficient of permeability equal to or better than 1×10^{-9} metres/sec. It is important that the material is of a uniform specification and it is laid in accordance with the specified method of working. The material should be protected from chemical attack and physical damage or distortion which could increase its permeability.

8.6 Ideally synthetic membranes should be flexible, durable, of very low gas permeability and exhibit high resistance to tearing or puncturing. They should also be inert to chemical or biological attack.

8.7 Whilst at many sites the base will require sealing for leachate containment purposes, because of the potential for gas to migrate in any direction, the need to install a gas barrier under the wastes should also be evaluated.

Closed sites

8.8 Barriers can be used around closed sites, but their application is limited by the depth of trench that can be dug and the fact that no barrier can be installed across the base of the site. These restrictions, coupled with the problems described above, mean that barriers should not be used as a sole means of control for closed sites.

8.9 For shallow sites with depths to about 5 metres a synthetic membrane can be laid in a trench to provide a barrier. They can be installed in 0.5 metre wide trenches, which should be dug to a depth greater than the maximum site depth. Such trenches should be laid just outside the waste with the membrane on the side of the trench remote from the landfill. Where there is no room between the waste and site boundary, construction of the trench can be at the waste edge.

8.10 A slurry trench can be used for depths greater than 5 metres or in difficult ground. A slurry is usually composed of bentonite and cement. The bentonite is hydrated and blended with cement prior to being fed into the trench. The trench, usually at least 0.6 metres wide, is filled with the slurry as the excavation proceeds, such that the slurry supports the sides of the trench. The mixture is self-setting with a consistency of stiff clay and a coefficient of permeability to water better than 10^{-9} metres per sec. Usually the trench is dug in soil and the soil discarded; however in some types of ground the soil itself may be used in the mix, provided a less reliable barrier does not result.

8.11 Slurry trenches can be dug to depths of 30 metres with specialised equipment, though trenches of that depth are not known to have been used in this country. Depths of 15 metres can be obtained with long reach excavation equipment. Installation of such trenches requires specialist skills, especially where variations of depth are necessary, the barrier is in very dry ground, or where the trench has to be keyed into underlying strata.

8.12 A method of improving the effectiveness of a slurry trench is by the addition of an HDPE (High Density Polyethylene) or other synthetic sheet within the slurry. Sections of sheeting may be lowered into the trench by a rigid framework and sealed to adjoining sheets with a special device. The effectiveness of this system relies on well-supervised installation, for it is only at this stage that measures can be taken to detect problems such as inadequate excavation. Failure to key into underlying strata will mean the barrier could be ineffective.

8.13 Grout curtains have been used to act as a gas barrier. These are usually constructed by drilling boreholes close together in a staggered pattern along a line. The interval between boreholes is dependent on the type of strata, though 1 metre centres is typical. A cement slurry or similar grout is injected under high pressure into each borehole to form the curtain. The effectiveness of grout curtains has yet to be confirmed and cannot be relied upon as the sole means of preventing gas migration.

Permeable trenches

Operational and proposed sites

8.14 Vent trenches, about 1 metre wide, filled with "no fines" crushed aggregate of uniform size can provide a route through which gas can vent. The side of the trench furthest from the landfilled waste should be sealed with a low-permeability barrier of natural or synthetic material and the rest of the trench lined with a fabric filter to prevent blinding of the medium. Perforated or slotted pipes of suitable strength material (such as class 3 HDPE, medium density polyethylene (MDPE), polypropylene or uPVC) should be installed in the trench and connected to surface vent pipes of similar construction. Such pipes should be at least 100 mm diameter. Vent pipe spacing should be determined from monitoring and site investigation data, but should generally not be greater than 50 metres apart. Adequate protection to the vent aperture should be provided to guard against ignition of the issuing gas and vandalism. The height of the vent pipe will be determined by local circumstances taking into account matters such as amenity and visual intrusion. Trenches should be located between the waste fill and the gas barrier or side of the site (see Figure 8.1.) and they should extend to the full depth of the wastes. For operational sites they can be built up as infilling progresses or they can be installed when landfilling is complete. Re-excavation to install these trenches can only be satisfactorily carried out where shallow sites, 8 metres or less in depth, are being treated. Trenches should be vented and then capped to prevent surface water ingress.

8.15 There are several difficulties in using permeable trenches:–

a) wind blown waste or fines can block or partially block trenches, reducing gas flow;

b) emissions of gas can create an odour problem before the site is completed and may become a nuisance to local residents or to site operatives; and

c) the trench can form a drain for surface water or leachate into the base of the site. Impermeable barriers may therefore be necessary to prevent water ingress.

Closed sites

8.16 Excavation to install these trenches at closed sites can only be satisfactorily carried out where shallow sites, 8 metres or less in depth, are being treated.

Gas drains

8.17 Gas drains installed within the landfill as filling progresses provide an additional route to improve the effectiveness of gas wells.

8.18 Gas drains of about 1 metre square cross-section, constructed from no-fines crushed aggregate, stones or similar material can be laid horizontally in each layer of the waste in a lattice fashion as filling proceeds and connected to the surface via the wells. Perforated pipes should be installed in the fill material to assist gas flow. The drains should be installed by shallow excavation of the most recently tipped layer of waste backfilling with aggregate. For convenience they can be installed at 6 to 12 metre vertical intervals (the pipes being usually supplied in 6 metre lengths), and extend 5 to 10 metres horizontally from the wells.

Gas wells

8.19 These are similar in design to boreholes. They are normally constructed vertically using slotted or perforated class 3 HDPE, MDPE or polypropylene pipe up to 225 mm in diameter, surrounded by suitably sized no-fines crushed aggregate. In some instances fabric filter linings have been used to prevent entry of fines into the aggregate which could reduce the efficiency of the well. Collection pipes should be connected to a vent which is sealed at the surface with bentonite

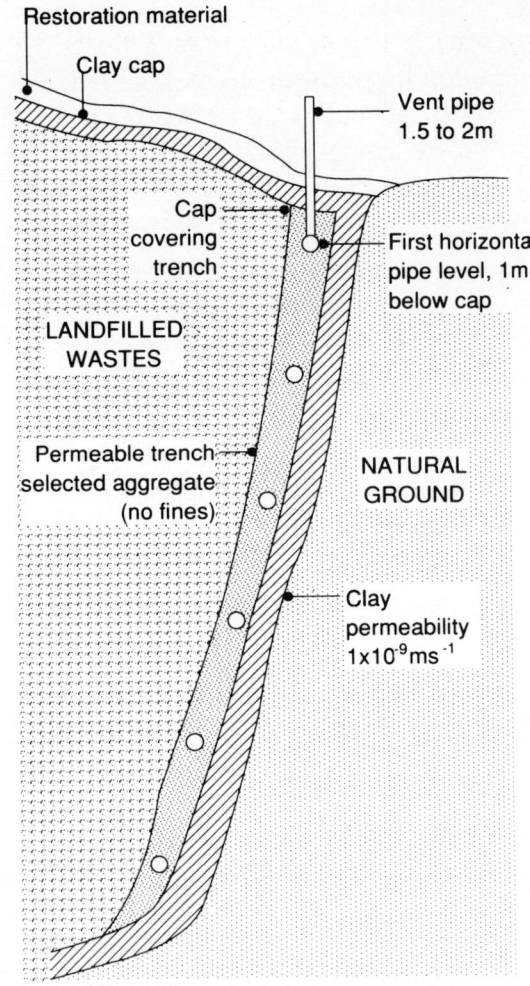

Figure 8.1b Permeable trench venting system, section A—A

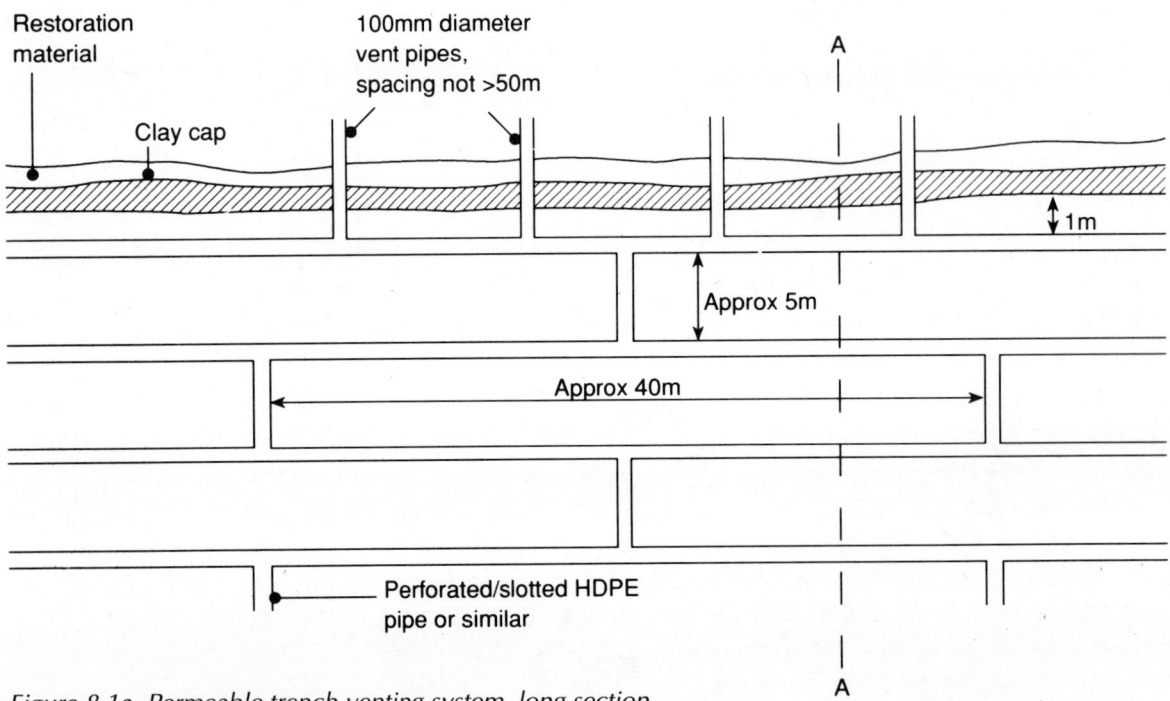

Figure 8.1a Permeable trench venting system, long section

or similar material. The effectiveness of the well will depend on the permeability of the surrounding wastes. The wells can be constructed on a grid system as filling proceeds although steps will normally need to be taken to prevent accidental damage during the site operations. At closed sites, wells should be drilled into the wastes. For passive venting of sites the wells should generally be constructed about 10 to 15 metres in from the edge of the wastes and not more than 20 metres apart. Where the sides of the site are benched or sloping, additional wells may be needed further within the body of the wastes to intercept their full depth. Figure 8.2. shows a typical design of a well for passive venting.

8.20 There are various problems in designing wells which need to be considered. In some wells where concrete has been used, there has been severe corrosion of the concrete in a relatively short time. Wells have also been flooded by high levels of leachate and therefore in some circumstances may required to be pumped. Subsidence has caused wells to warp, crack or block making them unusable. Well heads set too deeply into the site are also not recommended because of the risks involved in climbing inside to carry out measurements for monitoring. Wells also need to be located easily and their positions should be surveyed and recorded on a plan of the site.

8.21 The main advantage of wells are that all levels of waste are intercepted. In addition they retain their integrity better than trenches constructed within the waste, which can become distorted as settlement occurs, or may be flooded by perched water.

Gas pumping systems

8.22 Peripheral wells and trenches may be pumped to increase their sphere of influence and hence the volume of gas extracted. This will be necessary at all deep landfill sites (ie, with depths in excess of 8 metres) during the most active phase of gas evolution and where monitoring has shown passive venting to be inadequate. To be effective, pumped wells will need to be spaced at less than 40 metre intervals around the site perimeter. In areas of high risk this spacing should be reduced. The installation of a second system closer to the centre of the site may be necessary to reduce gas pressure there. Figure 8.3 shows a typical gas extraction well.

8.23 Each well or trench will have different characteristics with respect to the amount of gas that can be extracted and therefore each will require a separate control valve and sampling point. Valves and sampling points should be easily accessible although whether they will be installed above or flush with ground level will depend on the location and after-use of the landfill.

Figure 8.2 Passive gas venting well

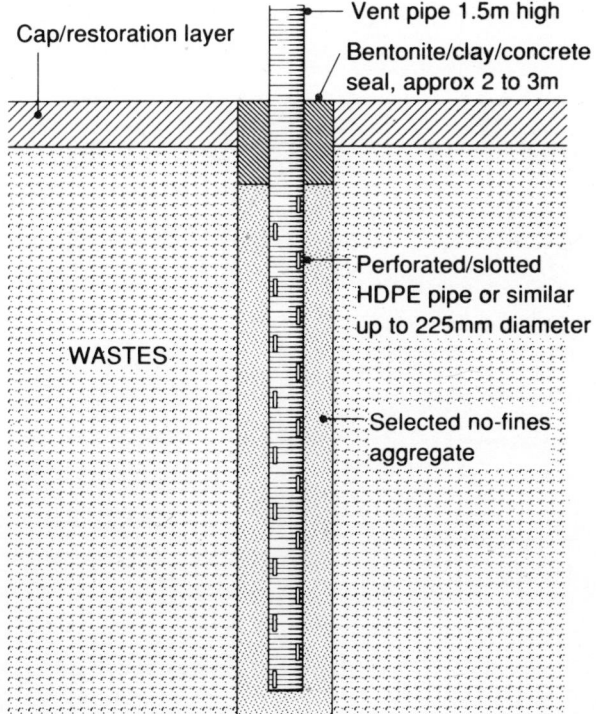

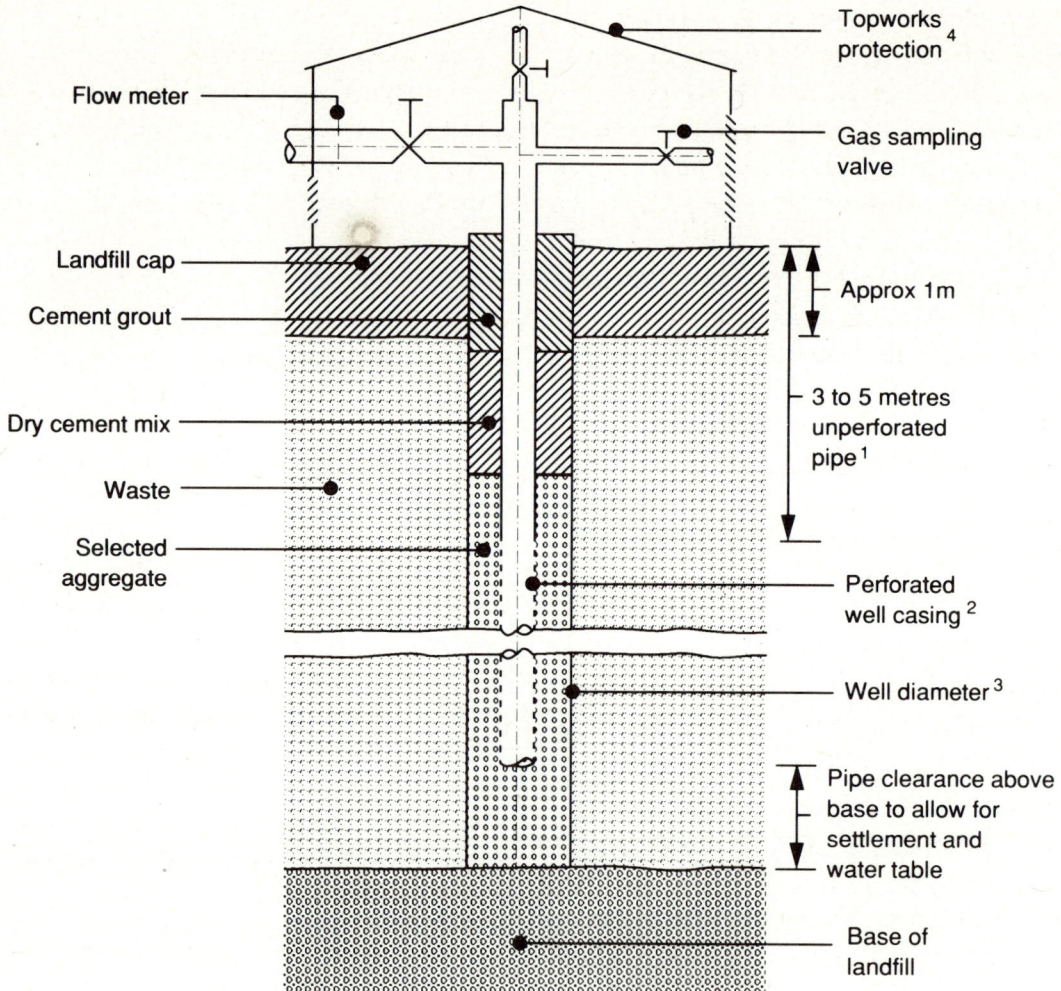

Notes

1. Exact length of unperforated pipe depends on waste depth and presence of water/leachate table.

2. Approximately 10% of pipe area should be perforated, at least 100mm diameter, UPVC, MDPE, HDPE or polypropylene.

3. Overall diameter determined by extraction rate required, established by static tests.

4. Topworks may be suitably protected above ground by a robust ventilated enclosure, or below ground in a manhole which does not compromise the integrity of the cap.

Figure 8.3 Typical landfill gas extraction well

8.24 It is important to monitor and control well pumping to ensure that over pumping does not occur, thus drawing excess air into the system. This is most likely when too few gas wells have been installed to prevent migration. The system should be designed and balanced to prevent gas migration without aerating the wastes. Overpumping leads to temporary cessation of landfill gas generation around the affected well. When pumping is stopped, gas production recommences after a short time and often more gas is evolved temporarily than the previous steady state condition.

8.25 It could be dangerous to pump landfill gas in its flammable range. To minimize the risk of explosion in the pipework and equipment, the composition of gas being pumped should be monitored for flammable gas/methane and/or oxygen concentrations or other steps taken to prevent danger from explosion. Any monitoring equipment should be installed close to the pump and linked to an alarm and pump shut-down system, set to operate at pre-determined safe levels. The equipment and system should be so designed as to ensure that venting can safely continue to maintain control of the gas.

Gas plant

8.26 Waste Management Paper No 26 provides information on the plant and equipment used in gas pumping systems. The following additional points on flarestacks and control systems should be noted.

8.27 Where the gas is not capable of supporting combustion, direct venting to atmosphere may be employed, although odour problems may need to be dealt with. Alternatively, support fuel may be needed to allow combustion and may also assist in reducing odour problems. In such circumstances it may be that in certain parts of the gas pumping system, flammable gas is mixed with air in the flammable range. Consideration needs then to be given as to what additional safeguards are necessary to prevent a dangerous occurrence due to an explosion within the gas plant.

8.28 The purpose of flaring gas is to dispose of the flammable constituents safely, and remove odour to prevent nuisance. However, consideration needs to be given to the possible health risks associated with the products of combustion, if the flarestack is located near to residential development. It is recommended, therefore, that the minimum flame temperature in such circumstances should be as high as possible and not less than 850° Celsius. For landfill sites where the extracted gases contain significant amounts of minor constituents whose combustion products could create a health risk, then flarestacks temperatures should be in excess of 1100° Celsius, with due consideration for the residence time.

8.29 Flarestacks should be properly maintained. There should be a sufficient supply of flammable gas, as flarestacks operating at less than 20% of their capacity may not burn at the correct temperatures.

8.30 All gas plant, including the pump, monitoring equipment and flarestack should be positioned in a secure location. Only properly designed plant should be used. To minimize the risk of propagation of an explosion in the pipework and equipment upstream, suitable flame arrestors should be provided. Ground radiation temperatures should be considered when calculating the height of the stack.

8.31 Flarestacks may need planning consent if they do not form part of an existing consent for a gas management system.

8.32 Where the potential for gas migration is significant, actively pumped gas migration control equipment should be fitted with an automatic flame ignition system within the flare stack and a 24 hour telemetry link with a manned service (eg, security service contractors). Such a system should be tripped on failure of the power system, or flame failure, and/or at preset alarm levels for adverse methane or oxygen concentrations of abstracted landfill gas. Emergency procedures, and notification to responsible officers for the plant, should be lodged with duty officers, and stand-by equipment for critical elements within the plant should be readily available. The advice of a specialist or equipment supplier should be sought with regard to the retention of spare parts.

8.33 Increasing sophistication of automated gas monitoring systems, linked for example to a series of gas monitoring boreholes, should allow for their incorporation into telemetry links to central offices. This is particularly important at sensitive sites with adjacent development where immediate indications of adverse conditions are required. Telemetry links from alarm systems within properties should also be provided, where possible.

Valves and valve selection

8.34 When selecting valves, the corrosive properties of the gas should be considered together with the temperature conditions likely at an exposed site in winter. PVC valves are prone to failure at low temperatures. Therefore lined metal, or HDPE or certain proprietary valves are preferable in these conditions.

8.35 Wellheads and wellhead valvegear should be "tamperproof" and protected from vandalism or accidental damage by lockable covers with either removable or lockable valve handles.

8.36 Gas concentrations in the wells should be checked regularly and the valve settings and flow rates adjusted where necessary to maintain the system in balance.

Connecting pipework

8.37 The pipework connecting the well heads or vents to the pumping system should be of thermoplastic material such as HDPE, MDPE, polypropylene or similar material. Pipe diameters will be determined by the gas flow rate and the need to maintain minimum pressure drops. Typically pipes are between 100 mm and 150 mm diameter between well heads and 150 to 200 mm diameter for lines to pumps. It is preferable to lay the pipework below ground. Above-ground, pipes need protection from direct sunlight and are only acceptable on a temporary basis. Exposed pipes should also be protected from freezing. Pipe joints should be kept to the minimum, be well sealed and flexible enough to compensate for movement caused by settlement and temperature variations.

8.38 Condensate forms when the warm, moist gas is drawn through the cooler pipework. Condensate traps should be incorporated in the pipework, prior to the pumps, to prevent blockages. Pipework should be laid to a fall to allow condensate to flow to the traps. Where low points exist within pipework additional traps may be needed. The corrosive properties of condensate should be taken into account in the design of the discharge from such traps.

Maintenance of gas pumping systems

8.39 Pumping systems operate in a hostile environment, handle a variable and corrosive gas and so should have a high reliability. It is vital therefore that a programme of planned maintenance is adopted covering all aspects of the system for as long as the gas control is needed at the landfill. Maintenance procedures or contracts should cater for the following points:–

(a) Wells and interconnecting pipework installed in the landfill will be subject to stresses caused by settlement and blinding by fines drawn to the intake by the passage of gas;

(b) Pumps and motors will be under constant operation and will need pre-planned maintenance (eg. bearings, brushes);

(c) Metal pipe fittings and condensate traps will be subject to corrosion;

(d) Cleaning (de-silting) of flame and flashback arresters;

(e) Checks for possible embrittlement of any exposed PVC pipework;

(f) Automatic alarm and telemetry systems will need planned maintenance possibly by the supplier to ensure highest reliability.

8.40 Some or all of a gas pumping system may need renewal during its period of use. When carrying out major maintenance, consideration should be given to incorporating newly developed measures to upgrade or improve the reliability or instrumentation of the system.

Effects of site operations on gas management systems

8.41 The effect of perched water has been described in Paragraph 5.3. This can be prevented by use of permeable cover or the construction of drains to interconnect successive waste layers and, where possible, the removal of low permeability intermediate cover. The incorporation of temporary site roads using hard-core or similar permeable material may also create pathways for gas movement within the wastes.

8.42 The mixture of gypsum wastes with biodegradable wastes can lead to the evolution of hydrogen sulphide (see paragraph 3.2). It is therefore recommended that large proportions of gypsum wastes are not co-deposited with household, commercial or similar wastes.

Settlement

8.43 At some sites where active gas control systems have been installed, high initial rates of settlement have occurred after starting gas extraction. For example at a 40 metre deep site, 18 months after completion of a phase there was settlement of about 0.5 metres, and in the next 6 months, after gas extraction had started, settlement in the same phase was about 1 metre. Additional settlement can occur from the loss of carbon in gas evolution.

8.44 The pipework fittings on well-heads should be designed to minimize settlement induced damage and reduce any loss of efficiency. Where well heads are located below ground level, the chamber surrounding the well-head should be constructed from a durable lightweight material such as glass reinforced plastic or galvanized steel.

Safety on sites

8.45 Operators should provide suitable training and instruction to employees who work permanently or transiently on site. The training should enable employees to understand how dangers can arise from landfill gas and how they may be avoided. The operator should also ensure that any contractor working on site is also informed of the hazards and the necessary precautions. No work should be carried out unless the risk to health has been assessed. The Control of Substances Hazardous to Health Regulations 1988 require that the exposure of employees to hazardous substances must be prevented or, if this is not reasonably practicable, adequately controlled. This should be done, so far as is reasonably practicable, by measures other than the provision of personal protective equipment. Monitoring of exposure and health surveillance may be required. The HSE booklet "COSHH Assessments" provides information on this topic.

8.46 Smoking should be prohibited on site. This should be enforced for all employees and visitors. There should be prominent warning notices in appropriate positions near the entrance to the site. Any fire on the site should be treated as an emergency and extinguished immediately.

8.47 Electrical equipment which is necessarily located in areas where accumulations of flammable gas could occur, should be selected, installed, and maintained in accordance with the requirements of BS 5345. Portable equipment should be similarly treated eg telephones, site radios etc.

8.48 Buildings and any other enclosed structure on site should be designed and built to prevent, so far as is practicable, accumulation of flammable gas inside them. Adequate circulation of fresh air will generally be required. Buildings on site should be regularly monitored for the presence of flammable gas and fitted with alarms set at a maximum of 20% LEL for methane. Air spaces under temporary cabins should also be well ventilated and may require monitoring at points where gas is likely to accumulate, eg. adjacent to service duct entries.

8.49 Unnecessary creation of enclosed spaces on site, eg. by inversion of a skip for maintenance, should generally be avoided. Vehicles should be parked in well ventilated positions to prevent accumulations of gas within them. Vehicle engines have been known to "race" as a result of methane entering engine intakes. Vehicles and plant may need to be banned from areas where particularly high gas emissions are found.

8.50 Instructions should be issued to all employees that no-one should enter any confined space below ground level, such as culverts and manholes, where there is poor ventilation, until an authorised person has clearly ascertained that it is safe to do so. A safe system of work should be established in accordance with HSE Guidance Note GS5, entitled "Entry into Confined Spaces". This should involve a permit to work system, and rescue procedures should be rehearsed. Entry into confined spaces where there is a flammable atmosphere should be prohibited.

8.51 Physical barriers should be in place to prevent unauthorised access to culverts and other confined spaces. Children have died when playing in culverts on landfill sites.

Gas management

8.52 The various control and monitoring strategies discussed in this and the preceding chapters should be brought together to provide an effective, permanent gas management system. Generally this is likely to require the installation of more than one means of gas control. As the effectiveness of many systems reduces with time, an element of over-design is necessary and is a matter for specialised advice.

8.53 The gas control system for each phase to be restored should be designed and installed as part of the restoration. The system will need frequent monitoring and when gas pumping commences, gas wells may need frequent adjustment of the head valve to meet desired gas quality.

8.54 For sites where waste is being used for land-raising, it is recommended that permeable strata below the site should be protected from possible gas migration by use of a barrier. Adequate gas control can then be achieved by gas wells and drains. Where such sites are built on clay or similar low permeability strata, an artificial barrier may not be necessary.

8.55 In some areas it has been the practice of site operators to cap sites with more permeable material such as shales. Whilst this may allow venting from upper layers of the landfill it may not control gas migration from lower levels and can make leachate control difficult.

Commercial utilisation

8.56 Commercial utilisation has been discussed in Waste Management Paper No. 26. It is not the purpose of this paper to discuss the exploitation of landfill gas as a resource except to note that this a positive benefit of landfill waste disposal. Wherever practicable these opportunities should be taken up. Commercial utilisation may provide an option which has lower net costs than that of gas control systems alone. This may have a beneficial effect on gas control by reducing gas pressure within the site and assist in covering the cost of maintenance necessary with a pumped gas system. Gas control systems should be capable of independent operation from commercial utilisation schemes, and have capacity for the extra loading when the commercial scheme is not in use.

CHAPTER 9

Development on or around landfill sites

Introduction

9.1 Decisions on development are matters for the relevant planning authorities. The advice given in this Chapter serves to highlight the difficulties that may be encountered where development is proposed near to or on landfills. Whatever form of development is proposed on or adjacent to any landfill site, it should not compromise the pollution control measures. Therefore, it is essential that all prospective developers undertake a comprehensive investigation as part of any proposal.

Development on landfill sites

9.2 As well as providing a disposal route for a wide range of wastes, controlled landfill also offers the possibility of restoring derelict land and mineral extraction sites to a beneficial after-use. Whatever form this takes, the integrity of the cap and the leachate and gas management systems should be maintained for as long as they are required. In many instances a landfill may not become stabilised for periods well in excess of 15 years. Agricultural, public open space, recreation or conservation are therefore the most appropriate after-uses until the site has stabilised.

9.3 It is accepted that in many cases there may be pressures to introduce other forms of development before the landfill has stabilised. Many landfills, originally outside urban areas, have become prime locations as development has encroached upon them. In recent years the demand for former landfills to be used as building plots has increased. Department of the Environment Circulars 21/87 and 17/89 outline the measures needed to conduct a survey of a completed landfill to assess its suitability for development. The Inter-Departmental Committee on the Redevelopment of Contaminated Land Guidance Note 17/78 "Notes on the Development and After-use of Landfill Sites" and approved building Codes of Practice should also be consulted. Information on the steps to be taken during construction on landfills has also been presented in Waste Management Paper No 26. Developments such as supermarkets, warehouses and blocks of flats have taken place on sites which are still evolving gas. These developments have been undertaken with expert advice and proper precautionary measures incorporated in the design and construction stages. Equally there are examples of developments that have run into difficulty by failing to take all relevant factors into account. It is also difficult to ensure that protective measures are maintained in private housing. Such measures may be defeated or breached by the actions of occupiers outside the control of developers, landowners and local authorities. In addition, adequate provision of protective measures may not be made in outbuildings, extensions, garden sheds, greenhouses etc. Domestic housing should not therefore be built on landfills which are gassing or have the potential to produce significant quantities of gas.

Development of land adjacent to landfill sites

9.4 Where development is proposed within 250 metres of a landfill site, whether operational, awaiting restoration or restored, the developer will need to take account of the proximity of the proposed development to the landfill and investigate the geology and topography of the area. Under the Town and Country Planning General Development Order 1988, planning authorities are now required to consult Waste Disposal Authorities on development within 250 metres of a landfill site either active or closed within the last 30 years. Local authority registers of land which may be contaminated, compiled under S.143 of the Environmental Protection Act, will record the locations of both closed and operational landfills and will provide an additional aid to identify such sites. Whilst for operational sites, all gas should be controlled as provided in Paragraph 2.4, it is possible that some minor escape, which may originally have been acceptable, could become significant by a change in use of the land around the site.

9.5 For all sites where retrospective action has been taken to control gas or at closed sites where no gas control measures have been implemented, it is possible that migrating landfill gas has filled the pore space of the adjoining strata outside the site. In such circumstances the effectiveness of control measures should be examined at the site, and specialist advice should be obtained on whether the strata should be cleared of the gas by active pumping, or whether the strata should be left to vent naturally.

9.6 When no gas is found in strata, but there are significant quantities of gas within the site or there is the potential for evolution of large quantities of gas, consideration should be given to possible gas migration pathways between the site and the development, especially through underground services. The proposal for the development should include measures to prevent gas using these pathways.

9.7 It is recommended that where landfills are actively producing large volumes of gas, or have the potential to produce large quantities of gas, and housing development is proposed, that no house, garden shed, greenhouse or any domestic extension should be constructed within 50 metres of the boundary of the infilled wastes, and no garden should extend to within 10 metres of the wastes.

Highways on or adjacent to landfill sites

9.8 There are various risks associated with all roads running over or adjacent to landfill sites where quantities of gas are being evolved. Consideration needs to be given to the accumulation of gas in underground services, especially where there are possible ignition sources such as in lamp-posts. When roads are constructed over former landfill sites, specialist advice should be sought to ensure safe control of any gas that may be present. Care also needs to be taken in the feasibility, design and location of any space with public access where ventilation might be restricted eg pedestrian underpasses.

Conclusion

9.9 Whenever development is proposed on or adjacent to a landfill site, a comprehensive investigation of the site, the development, and the possible effect of the development on the landfill is essential, which should be supported by monitoring as described in Chapters 6 and 7.

APPENDIX A
Typical landfill gas composition

Component (%Volume)	Typical value (% Volume)	Observed Maximum (% Volume)
Methane	63.8[1]	88.0[2]
Carbon Dioxide	33.6[1]	89.3[1]
Oxygen	0.16[1]	20.9[1,3]
Nitrogen	2.4[1]	87.0[2,3]
Hydrogen	0.05[4]	21.1[1]
Carbon Monoxide	0.001[4]	0.09[2]
Ethane	0.005[4]	0.0139[2]
Ethene	0.018[4]	–
Acetaldehyde	0.005[4]	–
Propane	0.002[4]	0.0171[2]
Butanes	0.003[4]	0.023[1]
Helium	0.00005[4]	–
Higher Alkanes	<0.05[2]	0.07[1]
Unsaturated Hydrocarbons	0.009[1]	0.048[1]
Halogenated Compounds	0.00002[1]	0.032[1]
Hydrogen Sulphide	0.00002[1]	35.0[1]
Organosulphur Compounds	0.00001[1]	0.028[1]
Alcohols	0.00001[1]	0.127[1]
Others	0.00005[1]	0.023[1]

Notes:–

1. Data taken from Waste Management Paper No 26.
2. Published data supplied by Aspinwall & Company.
3. Entirely derived from the atmosphere.
4. Taken from: Guilani, A J "Application of conventional oil and gas drilling techniques to the production of gas from garbage", American Gas. Association Transmission Conference, Salt Lake City, Utah. 5–7 May 1980
5. Landfill gas is usually saturated with water vapour, up to 4% by weight, depending on the gas temperature. At 25°C a value of 1.8% by weight is typical.
6. When undertaking initial confirmatory analysis by gas chromatography, the first five compounds listed above are usually identified when looking for the presence of landfill gas.

APPENDIX B
Main relevant legislation

There are many statutes and statutory instruments which may be applicable to landfill gas monitoring, control and utilisation. Essential legislation is listed below, although the list may not be exhaustive. It should be noted that there are some differences in the statutes for Scotland and Northern Ireland in particular and reference should be made to these as appropriate.

A) Legislation relevant to operation, monitoring and control

Health and Safety at Work etc Act 1974
Health and Safety at Work etc (Northern Ireland) Order 1978

The Reporting of Injuries, Diseases And Dangerous Occurrences Regulations 1985 as amended
The Reporting of Injuries, Diseases And Dangerous Occurrences Regulations (Northern Ireland) 1986

Public Health Act 1936
Public Health (Ireland) Acts 1878–1907
Public Health (Scotland) Act 1897

The Control of Pollution Act 1974
Pollution Control and Local Government (Northern Ireland) Order 1978

The Environmental Protection Act 1990

The Town and Country Planning General Development Order 1988 SI.1988 (No.1813) as amended
The Town and Country Planning General Development Order 1973

Town and Country Planning Act 1990
Town and Country Planning Order 1991
Town and Country Planning (Scotland) Act 1972

Planning and Compensation Act 1991

Town and Country Planning (Assessment of Environmental Effects) Regulations 1988 SI.1988 (No.1199)
Planning (Assessment of Environmental Effects) Regulations (Northern Ireland) 1989 SRO No 20
The Environmental Assessment (Scotland) Regulations 1988 SI.1988 (No.1221) S.122

Control of Substances Hazardous to Health Regulations 1988 SI.1988 (No.1657) as amended

Control of Substances Hazardous to Health Regulations (Northern Ireland) 1990

B) Legislation relevant to landfill gas utilisation

Health and Safety at Work etc Act 1974

Health and Safety at Work etc (Northern Ireland) Order 1978

The Reporting of Injuries, Diseases and Dangerous Occurrences Regulations 1985 as amended

The Reporting of Injuries, Diseases and Dangerous Occurrences Regulations (Northern Ireland) 1986

Public Health Act 1936

Public Health (Ireland) Acts 1878–1907

Energy Act 1983

Gas Act 1986

Gas Act 1972

Pipelines Act 1962

Gas Quality Regulations 1983

Gas Metrication Regulations 1980

The Notification of Installations Handling Hazardous Substances Regulations 1982.

The Notification of Installations Handling Hazardous Substances Regulations (Northern Ireland) 1984.

Control of Substances Hazardous to Health Regulations 1988 SI.1988(No.1657) as amended

Control of Substances Hazardous to Health Regulations (Northern Ireland) 1990

APPENDIX C
Monitoring for landfill gas in buildings

Role of a nominated person

C.1 It is recommended that every authority and employer involved in the monitoring of landfill gas appoint a nominated person, with deputies, to be responsible for dealing with any emergency which may occur due to landfill gas. The exact duties of the nominated person will vary, but he should ensure before any incident has arisen that he liaises with British Gas Districts.

C.2 Upon receipt of an emergency call the nominated person should:–

C.2.1 Alert the local emergency officer for British Gas plc. and the local police. Both parties should be informed of the circumstances and location of the incident.

C.2.2 If required, arrange for further monitoring staff.

C.2.3 Inform the local Environmental Health Department, especially where housing is involved and there is a need to rehouse occupants in emergency accommodation. Consideration should also be given to the preparation of a statement in response to enquiries from local press, radio and television.

C.2.4 In accordance with the requirements of the Reporting of Injuries, Diseases, and Dangerous Occurrences Regulations 1985, notify the Health and Safety Executive as soon as possible. Confirmation on Form F2508 should be sent within 7 days whenever the following concentrations are found, if they are attributable to landfill gas:–

1. Flammable gas is found in buildings at concentrations in excess of 1% v/v.

2. Carbon Dioxide is found in buildings at concentrations in excess of 1.5% v/v.

3. Any other gas is found in concentrations in excess of its occupational exposure standard (10 minute reference period).

Training of field operatives

C.3 Training and instruction in the hazards of landfill gas, precautions needed, procedures to be followed and techniques and limitations of gas measurement and sampling equipment is essential and should be given to all personnel involved in these duties.

Operational procedures for landfill gas surveys in buildings

C.4 On arrival at a building the following procedures should be adopted:–

C.4.1 On arrival, note date time and reason for visit.

C.4.2 Knock on door (do not use doorbell) and obtain access.

C.4.3 Show form of identification and inform occupier of reason for visit, and explain what measurements need to be taken.

C.4.4 Before entering building make gas measurements at door for landfill gas. The evacuation procedure described in Section C.5 should be implemented if the following gas concentrations are exceeded:–

- 1% by volume (20% of L.E.L.) of flammable gas/methane/hydrogen

- 1.5% by volume of carbon dioxide.

C.4.5 All measurements should be carried out using an intrinsically safe instrument until it is shown there are no methane concentrations in the building in excess of 1% v/v (20% L.E.L.). Measurements should then be taken using a more sensitive detector such as a flame ionization instrument.

C.4.6 Ask the occupier if any unpleasant or unusual odours have been noticed. Note their apparent source. Make gas measurements of all parts of the property, as indicated in Form C1, where gas is likely to collect and record all readings and locations. No cellar or other confined space should be entered unless the precautions outlined in HSE Guidance Note GS5 are followed.

C.4.7 Whether gas is detected or not, inform the occupier of results of testing. If the evacuation procedure is not implemented, leave the occupier with a standard letter of explanation and a copy of the report.

C.4.8 Upon leaving, note time of departure.

C.4.9 If significant gas concentrations (either flammable gas or carbon dioxide) are detected, action should be taken in accordance with the procedures described below in Table C1. The time of any such find and the peak and steady concentrations of gas should be recorded.

C.4.10 If the building is unoccupied, as far as possible check air bricks, letter box, service entries eg. water cocks and sewage outlets, also near-ground gas concentrations next to likely service entry points. If gas is detected the nominated person should be contacted who should then attempt to find and advise the occupiers or owners. It is important that before anyone enters the building, measurements are taken for gas. If the building is to be left unattended, a warning of the possibility of danger from gas should be posted. If no gas is detected, a standard letter advising the occupiers should be posted through the letter box.

Evacuation procedure

C.5 When landfill gas concentrations are found within buildings in excess of 1% v/v of methane or 1.5% v/v of carbon dioxide the affected occupied areas should be evacuated using the following procedures:–

Methane in excess of 1% v/v

C.5.1 *Advise occupants:* Warn of the danger and possible risk to life and property; advise necessity of evacuation.

C.5.2 *Ventilate:* All affected parts of the building should be ventilated by opening as many doors and windows as possible in as short a time as possible, then the occupants of the building should be evacuated to a safe distance. In larger buildings only non-electrical fire alarms should be used.

C.5.3 *Isolate all ignition sources:* All fires and naked flames shoud be extinguished immediately. Mains or piped gas should be turned off at the isolation cock. The electricity should be isolated at the meter, but flammable gas concentrations should be checked at the meter prior to this. An intrinsically safe gas detector should be used and the meter should not be switched on or off if gas concentrations exceed 1% v/v (20% of L.E.L). In this event the area should be ventilated.

C.5.4 Where there are infirm persons, or for larger buildings, where it is not possible to carry out immediate evacuation, doors and windows of affected parts should be ventilated. Procedures described in Paragraph C.5.3 above should be followed, but this should be carried out without delay.

C.5.5 If the flammable gas concentrations exceed 2.5% v/v (50% L.E.L.), the building should normally be evacuated immediately, leaving all doors open. Advise the Police and arrange security for the building.

Carbon Dioxide in excess of 1.5%

C.5.6 *Advise occupants:* Warn of the danger and possible risks, advise necessity of evacuation.

C.5.7 *Ventilate:* Open as many doors and windows as possible in as short as possible a time, then evacuate the room or immediate area.

C.5.8 As a precaution all fires and naked flames should be extinguished.

C.6 After the actions described in the preceding paragraphs have been completed, the nominated person should be informed of the circumstances.

C.7 When the Police arrive they will take control of the incident. They should be informed of the nature of the incident and of the necessity to check gas concentrations in any unoccupied property where entry cannot be gained.

Identify source(s) and ingress of gas

C.8 Adjoining occupied buildings should also be checked using the procedures outlined above. Where any other building is found to be affected the appropriate procedures should be followed.

C.9 When all occupied buildings have been checked for gas, any unoccupied buildings nearest the affected one, including garages, garden sheds, outhouses etc. should be checked.

C.10 *Extreme care* should be taken when re-entering the affected building. Remote monitoring techniques should be used to ensure staff are not exposed to risk to their safety so far as is reasonably practicable. All measurements should be carried out using an intrinsically safe instrument until it is shown there are no methane concentrations in the building in excess of 1% v/v (20% L.E.L.). Measurements should then be taken using a more sensitive detector such as a flame ionization instrument to establish the point of ingress of the methane. Gas samples may also be necessary to establish the source of the gas.

C.11 If it is later found not to be landfill gas, the occupier and/or owner and the appropriate agency (such as British Gas plc) should be informed.

Institute control measures

C.12 Buildings must not be re-occupied until control measures have been instituted to prevent further ingress of gas, the buildings have been adequately ventilated, and monitoring has shown that there is no further ingress of gas.

C.13 The extent of control measures to be taken prior to re-occupation will be site specific. It may only be necessary to seal a gap in the brickwork of a wall, but in other rare circumstances more drastic action may be necessary. Consideration can be given to ventilating the ground around the affected building. Specialist advice should be sought. The Waste Regulation Authority should advise the police and the Environmental Health Department when it is satisfied that the building can be re-occupied.

Continuous monitoring

C.14 The extent of monitoring to be undertaken can only be determined by a full knowledge of the circumstances. When no control measures have been undertaken and buildings remain empty, daily monitoring should suffice. After control measures have been instituted, then the use of a proprietary continuous monitoring device is recommended. The devices should be set to alarm at a methane concentration of 1% v/v (20% of L.E.L) or carbon dioxide in excess of 1.5% v/v.

C.15 When the control measures are found to work, continuous monitoring devices should no longer be necessary. However it is recommended that buildings should continue to be monitored until the site is shown to be gas free. This will especially be the case for houses, where householders might unwittingly compromise the installed control measures.

Procedures where gas is found, but evacuation not necessary

C.16 If concentrations equal to or less than 1% v/v (20% of L.E.L.) of methane or 1.5% of carbon dioxide are found in buildings, the occupants should be informed and advised that further monitoring is necessary, but statements should not be made which might alarm the occupier unduly.

C.16.1 *Ventilation:* Affected rooms should be ventilated by opening doors and windows, when methane concentrations are in excess of 0.25% v/v (5% of L.E.L.) or carbon dioxide in excess of 0.5% v/v.

C.16.2 *Sources of Ignition:* All sources of ignition should be removed from any area affected by methane.

C.16.3 *Identify Source of Gas:* A sample of methane should be obtained to assist in later analysis to identify the source of the gas.

C.16.4 *Identify points of ingress:* When the concentrations of gas are below those specified in paragraph C.16.1 within the affected rooms, it should be possible to close the windows and doors. The portable instruments should then be used to ascertain any increase in concentration of gas and to identify the point or points of ingress of gas. If the gas concentrations do rise then the occupant should be advised to ventilate the building continuously. Arrangements should then be made with the nominated person to install continuous monitoring devices alarmed to warn of concentrations at 1% v/v (20% of L.E.L.). Continuous monitoring devices should be checked daily.

C.17 Where it is clear that the gas is from a source other than landfill gas, the occupier should be informed of the need to take appropriate action. When the source of gas cannot be so easily traced then it should be assumed that the source is from landfill gas, until positively proved otherwise. The nominated person should be told of the circumstances and inform the appropriate statutory bodies (eg. WRA, Environmental Health Authority, public gas supplier, British Coal).

C.18 The affected rooms should be continuously monitored until it is evident that the gas concentrations are below those specified in paragraph C.16.1. The affected areas should then continue to be ventilated and gas measurements of adjacent buildings should be undertaken, giving priority to occupied buildings followed by any other empty buildings nearest the affected building, including garages, garden sheds, outhouses etc.

Institute control measures

C.19 Monitoring must continue until suitable measures have been instituted and shown to prevent further ingress of gas. The police and the Environmental Health Department should be advised when control measures have been successful.

Institute monitoring schemes

C.20 The extent of monitoring to be undertaken can only be determined by a full knowledge of the circumstances. When no control measures have been undertaken then the use of a proprietary continuous monitoring device is recommended. The device should be set to alarm at a methane concentration of 1% v/v (20% of L.E.L) or carbon dioxide in excess of 1.5% v/v.

C.21 When control measures have been shown to work, there will still be occasions where buildings should continue to be monitored. This will especially be the case at houses, where householders might unwittingly compromise the installed control measures.

Further action

C.22 When it is found that gas accumulations in buildings are caused by landfill gas then the interested parties should decide:–

a) Further monitoring requirements

b) Remedial control measures.

C.23 Because of the nature of landfill gas evolution and migration, it is essential that action is taken speedily to resolve problems, and the action being taken is subject to review, as the circumstances change.

C.24 Interested parties are likely to include: the site operator and/or the landowner, Environmental Health Authority, WRA, Police, relevant Insurance Company representatives, Health and Safety Executive, and the local public gas supplier. In certain places British Coal may also need to be consulted.

Table C1 Trigger Concentrations For Gas In Buildings

Gas Concentration	Location	Action
More Than:– 1% v/v Methane/Flammable gas or 1.5% Carbon Dioxide	General voids in occupied areas; unoccupied voids near occupied areas, points of ingress into occupied areas eg service ducts cracks at skirtings.	Evacuate building; ventilate building; control sources of ignition; identify source of gas; identify points of ingress; institute control measures; monitor continuously.
Equal to or less than:– 1% v/v Methane or 1.5% v/v Carbon Dioxide	Anywhere in the building eg occupied or unoccupied voids, points of ingress, service ducts, underfloor cavities etc.	Ventilate areas affected; control sources of ignition; identify source of gas; identify points of ingress; institute control measures; institute monitoring scheme.

FORM C1
Gas Measurement Survey in Building

ADDRESS OF BUILDING ....

....

CONSTRUCTION   DATE

                                                                             TIME

CAUSE OF VISIT   ....

SURVEY CARRIED OUT BY   ....

NAME OF PERSON SEEN   ....

NAME OF OWNER/OCCUPIER   ....

PROPERTY UNOCCUPIED/
ENTRY NOT OBTAINED   ....

DOORWAY CHECK BEFORE
ENTERING:

Flammable Gas   ....%LEL   Carbon Dioxide   ....v/v   Oxygen   ....v/v

LOCATION SURVEYED	CONCENTRATION			TIME
	FLAMMABLE GAS	CO_2	O_2	

LOCATIONS TO SURVEY:-

UNDER FLOOR SPACE	☐	CUPBOARDS	☐	LOFT	☐	WALL CAVITY	☐
UNDER STAIRS	☐	SKIRTING BOARDS	☐	CELLAR HEADS	☐	OTHER (SPECIFY)	☐

SERVICE POINTS:-

GAS	☐	ELECTRICITY	☐	TELEPHONE	☐	WATER	☐

DRAINS (Specify Number) ....

OUTSIDE GARAGE	☐	OUTBUILDINGS	☐	GREENHOUSE	☐	AIRBRICKS (No. ...)	☐

APPENDIX D
Description of monitoring equipment

Introduction

D.1 There are many types of portable instruments available for landfill gas monitoring. Care is needed in the use of all instruments as the gas may adversely affect their sensing elements. There are inherent limitations in each type of instrument and the conditions of use can also give difficulties in obtaining reliable results. Because of these limitations, additional samples should be periodically taken for confirmatory analysis by gas chromatography.

D.2 All monitoring should be undertaken by competent and trained operators who understand the limitations of the equipment being used.

Catalytic oxidation detectors

D.3 Instruments using catalytic sensing elements detect low concentrations of flammable gases as a percentage of the lower explosive limit (L.E.L.) of the gas. An instrument will respond to any flammable gas and therefore can be accurate only when measuring a pure sample of the particular gas for which it has been calibrated. If other flammable gases are present, there will be an error in the readings obtained.

D.4 These instruments require oxygen concentrations in excess of 12% by volume to ensure complete oxidation of the gas. At lower oxygen concentrations the instrument may not respond to the flammable gas and display a low or zero reading. To overcome this problem most manufacturers include a thermal conductivity device or oxygen concentration detector in conjunction with the catalytic circuitry.

D.5 The sensing agent of the catalytic instrument can deteriorate with age, and may be poisoned by the minor constituents of landfill gas. The instrument should therefore be calibrated using a standard gas mixture according to manufacturers instructions or at least once every 6 months, depending on usage, whichever is more frequent.

Thermal conductivity detectors

D.6 These can measure the total concentration of all flammable gases and carbon dioxide in the sample by comparing the thermal conductivity of the sample against an internal electronic standard representing normal atmospheric air. They can measure the full 0 to 100% range of gas concentrations, although the sensitivity at low concentrations, below the L.E.L. of flammable gas, is poor and insufficient to provide accurate results. With landfill gas the mixture of methane and carbon dioxide can cause response problems as each gas affects the thermal conductivity cell differently. Suppliers have recognized this problem and will calibrate their instruments using mixtures of the two gases. To obtain accuracy in readings it is essential that the sample being measured contains only the two gases for which the instrument has been calibrated. As the proportion of any other gas in the sample increases, so the error in readings increases. This type of detector measures a physical property of the sample gas, and is not affected by low oxygen concentrations.

D.7 A more recent development based on the thermal conductivity device is the binary gas analyzer. This instrument requires 2 measurements, one of the landfill gas and a second reading of the same gas with the carbon dioxide removed with an absorbing filter. The concentration of the methane and the carbon dioxide can then be easily calculated.

Combined catalytic and thermal conductivity detectors

D.8 Both catalytic and thermal conductivity detectors are incorporated in the same instrument and share a common display. It must be recognized that with two different methods of detection being used the "% L.E.L." scale is not a sensitive expansion of the "% volume" scale, and they are not interchangeable.

D.9 Both types of detectors require the sample gas to be drawn in a continuous stream. The volume of sample required may exceed the ability of the space being monitored to be recharged with the gas, if its volume is small, eg within a narrow bore shallow monitoring probe. In these circumstances air may be drawn into the space and non-representative readings obtained.

Infrared analyzers

D.10 Infrared analyzers can be used for measurement of specific components in landfill gas mixtures. They consist of an infrared source and detector. Specific analyzers are available for methane and carbon dioxide capable of measuring 0.5 parts per million (ppm) to 100% v/v gas. Infra-red analysers are relatively pressure sensitive and will not give accurate readings where a pressure differential exists between the reference and sample measuring cell.

D.11 The most recent design in the use of infrared analyzers is the infrared transmitter coupled to a receiver. The receiver can be mounted either in the same instrument or separately from the transmitter. This system can be used to detect the presence of gas through pathlengths of up to 700 metres. For short pathlengths the transmitter and receiver are housed together and the signal is reflected from a screen. These instruments measure concentrations of gas per metre of path and so, unless an array of separate analyzers is used, cannot distinguish an area of high concentration along the path amongst otherwise low concentrations. These units are suited for use in fixed monitoring stations.

Gas chromatography

D.12 Currently the most reliable method for analysing the major components of landfill gas is by GC. This technique should be employed for all permanent monitoring schemes to confirm measurements made by portable equipment. These instruments are rarely portable or rugged enough for use on a landfill site, but they can be set up in a field laboratory which may be most cost effective when large numbers of samples are to be analysed on a regular basis. GC is the only method of measuring the concentration of nitrogen in a sample.

D.13 As this method usually requires the employment of contract analysis facilities it is useful to confirm that the results obtained are accurate. It is possible to introduce errors in setting up the chromatograph, by using an incorrect standard for example, which would give rise to systematic errors in the readings obtained. Therefore, where readings appear to be greatly at variance with other readings it is advisable to undertake confirmatory analysis.

D.14 GC can also be combined with Mass Spectrometry (GC-MS) to analyse for trace components. This method is expensive and only need be employed where knowledge of these components is essential eg, for commercial utilisation of the gas or the identification of odorous compounds.

Flame ionization instruments

D.15 These instruments may be used to detect any flammable gas. In portable instruments they are normally used for measuring concentrations of 1-10,000 ppm, a common application being the detection of natural gas leaks. Such instruments typically employ a hydrogen/air flame and thus are not intrinsically safe. They must not be used, therefore, where flammable gas concentrations are in excess of 1% by volume (20% of L.E.L.). Also they are not able to operate in oxygen deficient atmospheres; operators should monitor the oxygen concentration before using them. Their accuracy is affected by the presence of gases other than methane, such as hydrogen, carbon dioxide and water vapour and some minor components of landfill gas. These instruments are most useful for detecting low concentrations of flammable gas at or above ground surfaces and for pinpointing sources of emissions. Owing to the likely conditions of measurement, the values reported should be treated with caution, providing a semi-quantitative assessment of the area being investigated.

Gas indicator tubes

D.16 These tubes provide a simple but crude indication for several components present in landfill gas and their use requires the least maintenance and operator training. The gas sample is drawn through a tube containing a reagent which reacts, producing a colour change, the amount of colour change corresponding to the concentration of the gas. However the potentially large number of minor components present in landfill gas could produce interference effects on the indicating reaction. These tubes should only be used as an indicator with little weight attached to the values obtained.

Oxygen monitors

D.17 Portable monitors are available for various ranges up to 25% of total gas. They are simple to operate but lose their sensitivity if used continuously with landfill gas due to moisture, corrosion, and poisoning problems. The electrochemical cell in the detector has a limited shelf life, and the instrument should, therefore, be regularly calibrated when in use. These instruments must always be used prior to operating flame ionization instruments in confined spaces and should be used to ensure that minimum oxygen concentrations are present for accurate determinations with other detectors. They should also be used to ensure safe conditions prior to entry into confined spaces on or near landfill sites.

Carbon dioxide monitors

D.18 Carbon dioxide can be detected and measured by infrared analyzers, thermal conductivity meters and gas indicator tubes. Solid state instruments are also available which can measure carbon dioxide concentrations in the range of 0.5 to 100% of total gas. An electro-chemical cell monitor is also available. Measurement of carbon dioxide is important when considering entry into confined spaces due to its asphyxiating properties.

Barometers & barographs

D.19 Gas pressure measurements in boreholes or wells should be made to assess the likelihood of gas migration and determine the effectiveness of the gas control system. Sensitivities down to 0.5 mm of water gauge are usually necessary. Atmospheric pressure should be recorded for the period during investigations and for about 48 hours prior to starting monitoring. Some correlations have been made between gas migration and changes in atmospheric pressure, although the actual effect on a specific site may be difficult to predict.

APPENDIX E
Sample collection and measurement techniques

E.1 Proper sampling is vital in obtaining reliable measurements of landfill gas. The major source of errors usually occur by dilution of the gas source, either because gas volumes present are very small or due to "leaks" in the sampling system. It must be recognized that many sampling points are small in volume and have low gas flow rates. Most instruments used for in-situ monitoring or for extraction of samples for subsequent analysis need a continuous flow of gas if stable average values are required. Most portable instruments contain manually operated aspirators typically with a volume of around 50 ml. In an unrestricted environment this volume may be passed through the detector within a few seconds. Thus in a minute as much as 300 ml of gas sample may have to be present and replenished from source if no dilution is to occur. Manual aspirators can be crudely controlled to reduce flow, or restrictions placed on or in sampling tubes connected to the instrument. Such measures may affect instrument calibration, which usually requires set flows across the detector. Other instruments such as flame ionisation detectors have in-built controlled flow rate devices. Any restriction such as a vacuum at the sampling point may reduce this flow rate which would produce inaccuracies in the reading obtained. Gas concentrations at the surface can vary significantly during measurement times due to climatic effects. In such circumstances it may only be possible to record average readings related to the duration of maximum and minimum values.

E.2 Sampling pipework should be flushed with the gas mixture prior to sample collection or measurement to avoid air dilution during sampling. When sampling volumes are small air dilution may be unavoidable, and so this possibility should be noted. After taking each sample the pipework should be flushed through with some of the next sample, before taking readings. Care must be taken to ensure that condensation or dust does not lead to blockages in sampling pipes or tubes. When taking field measurements with portable instruments, the equipment should have in-line filters installed, which are regularly checked, or filters should be provided in the sampling pipework as close to the instrument as practicable. Surface openings of vent pipes, boreholes and wells should be kept sealed prior to and during sampling.

E.3 When the volume of gas to be sampled is small in comparison to the amount of gas being drawn through the instrument, a peak reading will be obtained which will fall to a steady reading, proportional to the effectiveness of the seal of the sampling system and the volume of gas entering the sampling volume. Both the peak and steady concentration readings should be noted.

E.4 Instruments should be regularly calibrated and serviced according to manufacturers instructions. Records of services and calibration intervals should be kept.

E.5 When in use the instrument should be zeroed in the absence of landfill gas.

E.6 Landfill gas samples may be collected in a variety of containers which are listed in Table E1. It is important to be consistent in the methods of sampling, the apparatus used, and the analytical and measurement techniques employed, especially when comparing data taken from different sampling exercises. For most analyses of major components with concentrations exceeding 0.5%, material used for containment of samples is relatively unimportant. The possibility of adsorption of minor gas components of landfill gas onto the surface of sample containers should be recognized, and where necessary specially treated containers used to prevent this. Once collected, samples should be analysed as soon as possible. No container can be considered leak-proof and the usually wet and mildly corrosive gas may react with the container materials, resulting in degradation of the container and erroneous results.

Table E1 Sampling Containers

Container	Advantages	Disadvantages
Polyester/vinyl bags	Large capacity available—up to 44 litres.	Fragile and easily punctured.
Aluminium bags	Lightweight.	Body material reactive to certain minor components of landfill gas. Vacuum cleaning required. Water vapour condenses within bag and is difficult to remove.
"Teflon" bags	Large capacity available—up to 200 litres. Lightweight. Chemically inert.	Fragile Vacuum cleaning required Water vapour condenses inside bag. Possible puncturing of bag if sample removed by syringe.
Rubber bags	Large capacity available. Lightweight.	Fragile and easy to puncture. Water vapour condenses inside bag. Body material reactive to certain minor components of landfill gas.
Kevlar bags	Large size available. Lightweight. Relatively inert.	Fragile. Water vapour condenses inside bag. Vacuum cleaning required. Easily punctured if sample removed by syringe.
Stainless steel "bombs"	Robust. Good quality valves.* Pressurised versions available.	Relatively heavy. Body material reactive to certain minor components of landfill gas.
Glass	Relatively inert. Easily purged. Easy visual inspection.	Fragile. Sealing susceptible to leakage.
Copper	Robust. Capable of use with good quality valves.*	Body material reactive to certain minor components of landfill gas. Relatively small size. Relatively heavy.
Aluminium	Robust. Capable of use with good quality valves.* Pressurised versions available.	Body material reactive to certain minor components of landfill gas. Relatively heavy.

*"Good quality valves" are machine needle valves of either brass (with a stainless steel needle) or stainless steel

The above information was supplied by the Institute of Wastes Management

APPENDIX F
Other aids in gas monitoring and measurement

Introduction

F.1 This section considers the other methods which are available, or have been used to assist in detecting landfill gas or differentiating it from other sources. These methods are not routinely employed and may be expensive or unreliable (but not usually both).

Carbon dating

F.2 This method of analysis may be necessary if there is doubt about the source of the methane found. Carbon dating is based on the radio-activity of carbon-14, a naturally occurring isotope, which has a half-life period of 5730 years. Landfill gas has a different ratio of carbon-14 to carbon-12 than in either natural or mine gas. By using this technique it may be possible to distinguish landfill gas from the other gases, provided the sample is not mixed with gases of different "ages".

Use of stable isotopes

F.3 The ratio of stable (ie non-radioactive) isotopes of carbon ($^{13}C/^{12}C$) and hydrogen ($^{2}H/^{1}H$) can be used to characterize methane and any associated carbon dioxide. These ratios may be diagnostic of the source and have been used to differentiate geological methane (eg coal or natural gas) from biological methane (eg landfill or marsh gas). However because microbial processes use up methane containing the lighter ^{12}C isotope, the differentiation is more difficult if landfill gas is subject to methane oxidation or other microbial mediated processes involving methane after evolution.

Other measurements

F.4 In addition to portable gas monitoring equipment, other instrumentation may be of value in aiding interpretation, particularly with respect to in-situ monitoring. A simple vane may permit gas flow measurements or wind speed measurements to be made. Concentrated landfill gas emanating from boreholes or wells will have a density similar to air and flow measurements will indicate rates of evolution.

F.5 Temperature measurements may be useful indicators of waste activity within boreholes or reveal gas migration pathways beyond sites. High temperatures in migrating gas may reflect waste temperatures or be caused by methane oxidation.

F.6 Leachate/water levels should, where possible, be measured and liquid samples obtained for subsequent analysis. Gas migration may be both affected by liquid level changes and/or due to further microbial degradation of organic matter in migrating leachate.

Aerial photography

F.7 The taking of aerial photographs, both colour and infrared, can also assist in detecting the changes in vegetation due to migrating gas over an area around a landfill site. For the greatest benefit aerial photographs should be taken prior to the deposit of waste such that poor vegetation growth can be identified. Such photographs may show the extent of migration and therefore aid decisions on the areas where control measures are necessary.

F.8 The interpretation of results requires considerable skill. Due allowance should be made for other factors which may affect vegetation cover eg waterlogging, soil compaction and similar causes of crop stress that can present misleading information.

Aerial thermography

F.9 This is a technique using an infrared device similar to a camera to measure heat emission from the ground surface. A scanner is mounted in an aircraft or aerial balloon and can be fitted with a variety of lenses to vary the angle of vision. The scanner sends pictures to a video monitor, which can be recorded. The definition of the screen can be set to distinguish temperature differences of 0.2 degrees Celsius, with a black to white difference of 5 degrees Celsius. However this type of survey has to be carried out at night to avoid interference from the sun. The hot spots detected can be physically marked on the site during the survey, which can then be monitored with instruments. It should be noted that this technique does not detect gas emission, it only detects excess heat which could be caused by gas emission. In other respects this method shares similar disadvantages with aerial photography.

APPENDIX G
Gas measurements and sampling at boreholes & gas wells

Introduction

G.1 Gas measurements and sampling at monitoring boreholes and gas wells should be undertaken in a safe and consistent manner to enable fair comparison of the results obtained.

G.2 Training and instruction in the hazards of landfill gas, precautions needed, techniques and limitations of gas measurement and sampling should be given to all personnel involved in these duties.

G.3 It is essential that operatives follow the relevant advice contained in Appendices C, D and E when carrying out these procedures.

Safety

G.4 When specifying the materials to be used in the construction of the head of the borehole or gas well, special attention should be given to ensuring that the risk of creating friction or static sparks is kept to the minimum.

G.5 Pumped well heads or borehole heads should be locked or otherwise protected to prevent unauthorized tampering with gas valves and fittings and reduce the risks from vandalism.

G.6 The boreholes should be sealed to prevent gas escaping and air entering.

G.7 The borehole head should have a valve and suitable fittings to allow instruments and sampling bags to be directly coupled to the borehole without any escape of gas. The fittings should be so designed as to prevent water from lying within the sample path.

G.8 Operatives engaged in sampling and measurement should not smoke. Prior to opening up boreholes for sampling all cigarettes and naked lights within 5 metres of the borehole should be extinguished, and if in a public place "no smoking" signs should be displayed.

G.9 Only gas detectors constructed and tested to BS 6020 or certified to BASEEFA Certification Standard SFA 3007/1981 should be used, until it has been demonstrated that it is safe to use equipment not so certified.

G.10 Operatives should be provided with and wear suitable footwear and clothing during monitoring, including chemical resistant gloves to avoid contact with condensate, which can cause skin burns. Operatives should be warned not to "sniff" gas and to take steps to prevent inhalation of landfill gas whenever necessary.

G.11 Operators must not enter any confined space as part of borehole monitoring unless they follow the guidance provided in HSE Guidance Note GS5 entitled "Entry into Confined Spaces".

G.12 Adequate supervision and a safe system of work should be provided for operatives undertaking monitoring duties on their own.

G.13 No sampling of leachate should be undertaken at any gas abstraction well whilst the pumping system is operational.

G.14 If a well head of a gas abstraction system has to be opened to atmosphere, the well should be isolated or the pumping system switched off. To ensure the best control, it is recommended that this only be undertaken by written authorization of the site manager.

Equipment to be used

G.15 Operators should have the following equipment available for use. Additional equipment may also be necessary on occasions and that listed below will not always be necessary at every visit to a specific site.

G.15.1 A flammable gas detector for both high and low concentrations of gas.

G.15.2 A portable gas detector for measuring carbon dioxide for between 0 and 100% v/v of gas.

G.15.3 An oxygen monitor for concentrations up to about 25% v/v.

G.15.4 Gas sampling pump and suitable sampling containers.

G.15.5 A manometer for measuring gas pressure.

G.15.6 A suitable temperature probe for measuring gas temperature.

G.15.7 A barometer or barograph.

G.15.8 A metered plumb-line for measurement of leachate or water level in wells or boreholes.

Pre-sampling procedures

G.16 Office procedures should be established to ensure that all monitoring equipment is regularly checked. Gas detectors should be calibrated, and be within calibration date. Batteries and equipment filters should be in good condition, and sampling pathways checked for leaks. Flow rates on pumps should be checked, and squeezing aspirators examined for blockages.

Gas measurements

G.17 At each well or borehole the following should be adopted:–

G.17.1 Examine head of borehole or well. The appearance of the head should be checked for possible damage and any changes noted. The top should be removed and the fittings examined before any measurements are taken.

G.17.2 If the top of the well or borehole is damaged and the seal broken, measurements should be made around the head of the borehole, to ensure there are no dangerous accumulations of gas.

G.17.3 Measure pressure. The reference pressure inlet may need shielding from wind, and results should normally be noted in "Pascals".

G.17.4 Measure flammable gas concentrations, note peak and steady concentrations.

G.17.5 Measure oxygen concentrations, note peak (minimum or maximum) and steady concentrations.

G.17.6 Measure carbon dioxide concentrations, note peak and steady concentrations. Carbon dioxide and flammable gas concentration should preferably be measured at the same time with the two instruments connected in series. If this is done, the carbon dioxide monitor should be first in the gas pathway.

G.17.7 If gas sampling is to be undertaken, then after sampling, a second measurement of the flammable gas and carbon dioxide should be taken and noted as before. Where a steady reading cannot be achieved, the concentration should be recorded after a given time interval, depending on the sampling flow rate.

G.17.8 Any measurement of water depth in a borehole should be undertaken after completion of all gas measurement and sampling, or at a separate time. The head of the borehole will need to be removed for a depth probe to be inserted. This must not be done at a gas well if the gas extraction system is in operation and the well is not isolated.

Gas sampling procedure

G.18 The gas sampling pump and bag should be securely attached to the top of the well or borehole. The gas sampling equipment should be purged prior to taking the sample. Consideration needs to be given to the volume of the borehole, as well as the requirements of the analysis, when deciding on purging and sample volumes. Samples should be analysed as soon as possible after being taken. Gas chromatography (GC) has been recognized as the best available technique for analysis of landfill gas.

G.19 Gas sampling will only occasionally be necessary as a confirmatory guide to the measurements by portable instruments. Sampling should also be undertaken when unexpected results are obtained.

Reporting procedure

G.20 A reporting procedure should be established in the event of excessive gas concentrations being measured. The response will be site specific. As a guide at monitoring boreholes, urgent action should be taken when flammable gas exceeds 1% v/v, carbon dioxide from the landfill exceeds 1.5% v/v. If infra-red instruments are not used for measurement of flammable gas then oxygen concentrations below 19% v/v should also receive attention.

G.21 Form G1 provides an example of a recording form used for field measurements. It is recommended that a standard method be used for recording all gas sampling and measurement to ensure consistency.

G.22 Data-logging and computer-based recording systems may be usefully employed for field recording and subsequent data analysis.

FORM G1
GAS WELL OR BOREHOLE MONITORING

SITE .... DATE ....

OPERATIVE .... WEATHER ....

FLAMMABLE GAS INSTRUMENT Ser. No .... TEMPERATURE ....

CO^2 INSTRUMENT Ser. No .... WIND SPEED/DIRECTION ....

O^2 INSTRUMENT Ser.No .... BAROMETRIC PRESSURE ....

B/H Ref.	Flammable Gas	CO_2	O_2	Water Level	Pressure +/– mm/H_2O	Sampled Yes/No	Comment

1..

2..

3..

4..

5..

6..

Signed ....

Date ....

FOR OFFICE USE:

SEEN BY:

ACTION PROPOSED

Bibliography

Avoiding danger from underground services: HSE Booklet HS(G)47.

British Standard BS 5345: Code of Practice for Selection, Installation and Maintenance of Electrical Apparatus for Use in Potentially Explosive Atmospheres (Other than Mining Applications or Explosive Processing and Manufacture) (Parts 1-8, 1977–1989).

British Standard BS 6020: Instruments for the Detection of Combustible Gases (Parts 1–5 1981–1982).

COSHH Assessments. HSE Booklet 1988 ISBN 0 11 8854704.

Development of Contaminated Land. Joint Circular DOE 21/87– Welsh Office 22/87. ISBN 11 752025 X.

Entry into Confined Spaces: HSE Guidance Note GS 5.

Gas Safety (Installation and Use) Regulations 1984 as amended.

ICRCL 17/78 Notes on the development and after-use of landfill sites. Eighth Edition December 1990. Available from DoE Publication Sales Unit, Building 1, Victoria Road, Ruislip HA4 0NZ.

Industrial use of flammable gas detectors: HSE Guidance Note CS1.

Landfill Sites: Development Control. Joint Circular DOE17/89; Welsh Office 38/89. ISBN 0 11 752209 0.

Landfilling Wastes, Waste Management Paper No. 26, A Technical Memorandum For The Disposal of Wastes on Landfill Sites, Dept of the Environment 1986 HMSO ISBN 0 11 751891 3.

Measurement of Gas Emissions From Contaminated Land, D. Crowhurst, Dept. of the Environment Building Research Establishment, 1987, ISBN 0 85125 246 X.

Occupational Exposure Limits 1991: HSE Guidance Note EH40/91.

The Licensing of Waste Facilities, Waste Management Paper No 4, A Revision of Waste Management Paper No. 4 to provide a technical memorandum on the licensing of waste facilities including a review of relevant legislation. Her Majesty's Inspectorate of Pollution 1988, HMSO, ISBN O 11 752157 4.

The Town and Country Planning General Development Order 1988 SI.1988 (No.1813).

Monitoring of Landfill Gas. Institute of Wastes Management. 1990. ISBN 0 902944 18 5.

Glossary of terms

AEROBIC: In the presence of oxygen.

ANAEROBIC: In the absence of oxygen.

BIODEGRADATION: The breakdown of materials by the action of micro-organisms.

BIOLOGICALLY STABILIZED: The state where a system has completely degraded its nutrient source biologically to produce an inactive medium which is no longer capable of supporting growth.

BOD (Biochemical Oxygen Demand): A measure of the amount of material present in water which can be readily oxidised by micro-organisms and is thus a measure of the power of that material to take up the oxygen in water supplies.

BOREHOLE: A hole drilled outside wastes in order to obtain samples and to monitor for landfill gas migration. (See WELL.)

COD (Chemical Oxygen Demand): A measure of the total amount of chemically oxidizable material present in liquid.

"CONTROLLED LANDFILL": Where wastes are deposited in an orderly planned manner at a site licensed under the Control of Pollution Act 1974.

COVER: Material used to cover wastes deposited in landfills. Daily cover is used at the end of each working day to prevent odours, wind-blown litter, insect or rodent infestation. Final cover is the layer, or layers, of material placed on the surface of a landfill during its restoration.

EXOTHERMIC (Reaction): A chemical or biochemical reaction which results in the release of heat energy.

FLAMMABLE: A substance capable of supporting combustion in air.

GAS MIGRATION: The movement of gas from the wastes within a landfill site to adjoining strata, or emission into the atmosphere.

GAS MOVEMENT: The movement of gas solely within the wastes within a landfill.

IMPERMEABLE: Used to describe materials, natural or synthetic, which have the ability to resist the passage of fluid through them. It is usually expressed as the coefficient of permeability. This property is not absolute, and a cut-off coefficient of permeability of 10^{-9} m/sec for water is often used to describe a landfill liner material as impervious. The coefficients of permeability of materials for gases are likely to be greater.

INERT WASTE: Are strictly wastes that will not physically or chemically react or undergo biodegradation within the landfill environment.

LANDFILL: The engineered deposit of waste into or onto land in such a way that pollution or harm to the environment is minimized or prevented and, through restoration, to provide land which may be used for another purpose.

LEACHATE: The result of liquid seeping through a landfill and by so doing extracting substances from the deposited wastes.

"LEACHATE RECIRCULATION": The practice of abstracting leachate from the base of a site and returning it to the upper layers of the landfill.

L.E.L. (Lower Explosive Limit): The lowest percentage by volume of a mixture of flammable gas with air which will propagate an explosion in a confined space, at 25°C and atmospheric pressure.

L.F.L. (Lower Flammable Limit): The lowest percentage by volume of a mixture of flammable gas with air which will propagate a flame, at 25°C and atmospheric pressure.

METHANOGENIC: Methane producing microbial reaction.

MICROBE/MICROBIAL: Small organisms, usually single cells which are only visible under a microscope. They include algae, bacteria and fungi.

MONITORING: A process including physical examination, measurements by portable instruments, and analysis of samples to provide information for assessment of conditions.

ODOUR THRESHOLD VALUE: Lowest concentration of an odorous gas, which can be detected by sense of smell.

OPERATIONAL SITES: Sites which are not completely filled, operating under a licence, including sites which are temporarily closed for whatever reason.

PERCHED WATER: An accumulation of liquid at a level above that of the adjacent water table. Often caused by zones of low permeability strata (or wastes) which prevent downward percolation.

PERMEABILITY (Coefficient): A measure of the rate at which a fluid will pass through a medium. The coefficient of permeability of a given fluid is an expression of the rate of flow through unit area and thickness under unit differential pressure at a given temperature.

pH: A measure of the acidity or alkalinity of a liquid; a pH less than 7 is acidic, a pH greater than 7 is alkaline and a pH of 7 is neutral.

PUTRESCIBLE: A substance capable of being readily decomposed by bacterial action. Offensive odours usually occur as by-products of the decomposition.

U.E.L. (Upper Explosive Limit): The highest percentage by volume of a mixture of flammable gas with air which will propagate a flame, at 25°C and atmospheric pressure.

VENTING—ACTIVE: The removal, by forced extraction from wells or boreholes, of landfill gas. Usually from within a landfilled area of wastes.

VENTING—PASSIVE: The natural movement of gas from a landfilled area of wastes to atmosphere usually assisted by porous drainage media.

VENTING TRENCH: A trench containing porous granular material of uniform size which permits the free passage of gas.

WELL: A shaft installed in wastes for the monitoring and/or extraction of landfill gas. (See BOREHOLE.)